Tectonics
of Asia

Emile Argand
1879–1940

Portrait of Emile Argand. Reproduced from M. Lugeon, *Bull. Soc. Neuch. Sc. Nat.*, 1940, Vol. 65, plate facing p. 25.

Tectonics of Asia

Emile Argand

Translated and Edited by

Albert V. Carozzi
Professor of Geology
University of Illinois at Urbana-Champaign

Hafner Press
A Division of Macmillan Publishing Co., Inc.
NEW YORK
Collier Macmillan Publishers
LONDON

Copyright © 1977 by Hafner Press
A Division of Macmillan Publishing Co., Inc.

All rights reserved. No part of this book may be reproduced or transmitted in any form or by any means, electronic or mechanical, including photocopying, recording, or by any information storage and retrieval system, without permission in writing from the Publisher.

Hafner Press
A Division of Macmillan Publishing Co., Inc.
866 Third Avenue, New York, N.Y. 10022

Collier Macmillan Canada, Ltd.

Library of Congress Catalog Card Number: 76-14288

Printed in the United States of America

printing number

1 2 3 4 5 6 7 8 9 10

Library of Congress Cataloging in Publication Data

Argand, Emile, 1879-1940.
   Tectonics of Asia.

   Translation of La tectonique de l'Asie.
   Includes bibliographical references and index.
   1. Geology--Asia. 2. Carozzi, Albert V. III. Title.
QE289.A7313   1976   551.8'095   76-14288
ISBN 0-02-840390-8

*To my dear friend
Charles Eugène Wegmann,
Emeritus Professor of Geology
at the University of Neuchâtel,
with deep gratitude*

Contents

	List of Illustrations	ix
	Preface	xi
	Editor's Introduction	xiii
	Introductory Comments: Orogenic Cycles	1
Section I.	Art of Interpreting Structural Facts	2
Section II.	Tectonic Map of Eurasia	10
Section III.	Short Description of Precambrian and Caledonian Asia	12
Section IV.	Short Description of Hercynian Asia	22
Section V.	General Features of the Alpine Cycle	24
Section VI.	Behavior of the Tethys under Compression	30
Section VII.	Behavior of the Chains of the Tethys	34
Section VIII.	Types of Virgations	37
Section IX.	Deformation of Old Eurasia	43
Section X.	Cover Foldings, Basement Foldings, Reactivation	45
Section XI.	Duel between Indo-Africa and Eurasia	53
Section XII.	Tectonic Segmentation	57
Section XIII.	Segment of Central Asia	59
Section XIV.	Reactivation of Eurasia during the Alpine Cycle	61
Section XV.	Serindia	63
Section XVI.	Northern Periphery of Serindia	66
Section XVII.	Turanian Virgations	69
Section XVIII.	Evaluation of the Serindian Question	75
Section XIX.	Tien Shan, Kunlun, Nan Shan	80
Section XX.	Sino-Siberian Space	82
Section XXI.	Basement Folds of the Segment of Central Asia	86
Section XXII.	The Urals	88
Section XXIII.	Geguli-Ergheni Arc, Russian Platform, Caucasus	90
Section XXIV.	Basement Folds of Europe	92
Section XXV.	New Generalities on Basement Folding	102
Section XXVI.	Continuum and Discontinuum in Tectonics	118
Section XXVII.	Concrete Tectonics and Theories of the Earth	124
Section XXVIII.	Problem of the Pacific Ocean, Fixism, Mobilism	128
Section XXIX.	Glances at the Atlantic	138
Section XXX.	Glances at East Asia	158
Section XXXI.	Conclusion: Mobilism and Geologic Reality	160
	Epilogue: Asia	165
	Illustrations	166
	Notes	204
	Analytical Index	209

List of Illustrations

Portrait of Emile Argand (1879–1940) Frontispiece — ii

Editor's Introduction

Figure 1. Photograph of a Blackboard Sketch Drawn by Argand for His Lectures on Geotectonics, Dated October 31, 1925 — xv

Figure 2. Argand at Work on the Second Version of His Tectonic Map of Eurasia, Neuchâtel, 1920 — xviii

Figure 3. Last Page of Argand's Manuscript of *La Tectonique de l'Asie* — xx

Tectonics of Asia

Figure 1. Simple Virgation of the First Type — 167
Figure 2. Double Virgation of the First Type — 167
Figure 3. Simple Virgation of the Second Type — 167
Figure 4. Double Virgation of the Second Type — 167
Figure 5. Basement Folds — 169
Figure 6. Basement Folds of the Gondwana Continent — 171
Figure 7. Very Ancient Asiatic Drifts — 171
Figure 8. Schematic Tectonic Map of Eurasia — 173
Figure 9. Deformation Regime of Asia during the Alpine Cycle — 175
Figure 10. Plasticity of Asia — 177
Figure 11. Major Axial Behaviors of the Alpine Basement Folds of Eurasia (Western Half) — 179
Figure 12. Major Axial Behaviors of the Alpine Basement Folds of Eurasia (Eastern Half) — 181
Figures 13 to 18. Transverse Cross-Sections of the Zone of Confrontation Eurasia-Gondwana with the System Arisen from the Tethys — 183
Figures 19A and 19B. The Alps and Africa before and after the Great Distensions — 187
Figure 19C. Major Details of the Alps — 187
Figures 20A and 20B. General Features of the Mediterranean Frame during Alpine Precursor Times — 189
Figure 21. The Mediterranean Frame after the Major Activity of the Great Tertiary Paroxysms and before the Great Disjunctions — 191
Figures 22 and 23. The Mediterranean Frame during the Great Distensions and Disjunctions — 193
Figure 24. Sketch of a Post-Helvetian and Pre-Plaisancian State of the Mediterranean Frame — 195

LIST OF ILLUSTRATIONS

Figure 25. Sketch of a Quaternary State of the Mediterranean Frame 197
Figure 26. The Mediterranean Frame after the Major Activity of the Great Tertiary Paroxysms and before the Great Disjunctions 199
Figure 27. Present State of the Mediterranean Frame 201

Preface

La Tectonique de l'Asie, considered as the demonstration of Wegener's theory of continental drift, had a profound and lasting influence on European geological thinking. For instance, in the forties most of Argand's new concepts were taught at universities at the freshman level. This was the first time that I read his work. Later I climbed the same peaks and admired the same geological panoramas of the Alps that Argand had.

To read and to understand Argand requires a long and sustained intellectual effort, which is however rewarded by the enjoyment of a masterpiece that continually reveals unexpected facets of its perfection. I feel that the appreciation of Argand's fundamental contribution should be shared by as many geologists as possible; hence its translation into English.

La Tectonique de l'Asie is the text of an address. Accordingly, references are kept to a minimum and subdivisions are devoid of titles. I have checked the references for accuracy and presented them in full; I have added to the subdivisions short headings consisting of keywords. By now most of Argand's terminology has found its way into glossaries and textbooks of structural geology; therefore, my footnotes are limited to comments on unusual terms or expressions. The spelling of East Asian place-names has also been made to conform to modern practice. Argand's presentation is structured to such a degree that the analytical table of contents at the end effectively replaces a standard index.

In order faithfully to convey Argand's scientific message as well as his poetic style, I have striven to stay as close as possible to his original text, without being literal or trying to condense sentences that might at first glance appear rather redundant. In fact, there is no redundancy in Argand; everything is concision. Any attempt at simplification might lead to a complete misunderstanding of his thinking. And misunderstanding might likewise arise from the failure to uncover the particular sense Argand gave to a single word. When confronted with his unique word usages, I often turned to the provincialism of the language used in the French-speaking part of Switzerland, and putting myself back in the time of my Alpine training, I eventually discovered the intended sense of an unusual word. Naturally, it is not possible to convey in English all the subtle nuances of Argand's rich poetic style—poetry cannot be translated—but I hope that any betrayal on my part occurs only in minor matters.

While Argand had the genius to interpret from the mere compilation of pertinent articles and maps the geological structures of remote areas he had never seen, I had the chance to visit some of the areas that he discussed and described as the "light parade of the Oceanides" or the "ponderous basement folds of the Andes." When I compared what Argand called the concrete geological reality with his vision, I was very often able to reach a better understanding of both the reality and the vision.

This translation has been on my mind since I left the Alps for the Plains of North America many years ago. It was finally undertaken under the gentle advice and encouragement of my dear friend Charles Eugène Wegmann, once assistant to, and then successor of, Argand at the University of Neuchâtel, who provided me, through an extensive correspondence, with so many unpublished data on Argand that they themselves could be the subject of a book. Although I have liberally used this treasure of new information on Argand, I have respected the request of confidentiality pertaining to some of the documents in my possession. I am deeply grateful to Professor Wegmann for this unusual opportunity to reach a more profound understanding of Argand's personality.

Professor Jean P. Schaer of the University of Neuchâtel kindly provided me with photographs and other documents that illustrate many aspects of Argand's activity, as did Dr. Edouard Lanterno, curator of geology at the Museum of Natural History in Geneva. I am delighted to acknowledge these two important contributions. Mrs. Harriet W. Smith, geology librarian, and Mr. David A. Cobb, geography librarian, both of the University of Illinois, helped me, as generously as always, in the search for rare publications, unusual references, and remote place-names.

The accurate editorial work of Madeleine Sann Birns which led to numerous improvements is gratefully acknowledged. I am again deeply indebted to my wife for her typing, her devotion in proofreading, and her constant help, at all stages of this work, in my striving for perfection.

Albert V. Carozzi

Urbana, Geneva, Belém,
Salvador, Rio de Janeiro,
Lima, and Manila.
August 1976

Editor's Introduction

The time was August 10, 1922; the place was the opening session of the XIIIth International Geological Congress in Brussels; the room was very long and narrow, its acoustics poor. Emile Argand, professor of geology at the University of Neuchâtel—wearing a frock coat and a top hat, in full summer heat—entered the meeting place to deliver the inaugural address of the congress.* This address, *La Tectonique de l'Asie,* which lasted several hours, was a unique synthesis of global tectonics in the light of Wegener's hypothesis of continental drift. It was the manifesto of "mobilism" in geology, the revolution against the paralyzing and classical "fixism," which for so many years had obstructed any form of meaningful progress in geology.

The pace of Argand's presentation was breathtaking; his attempt at showing the Earth being built by a continuum of plastic deformation was bold, if not sublime. Dynamic tectonics was born while Argand was walking back and forth in front of his audience, often appearing to speak for himself rather than for the benefit of his astounded listeners. In fact, the presentation was the description of his tectonic map of Asia, affixed to the wall but barely visible.

The address was not an oratorical success. The subject was altogether too encompassing, and Argand's language was too full of technical terms, either newly created or borrowed from architecture, medicine, and the military arts. The fact that only a minority of the international audience understood French did not help matters. Nevertheless, the audience was highly enthusiastic as if they had intuitively grasped, if not really understood, the significance of this unusual talk. At the end of the presentation, Argand was besieged by geologists from all countries who wanted to comprehend fully the momentous conclusions he had reached on the structure of their respective homelands.

Argand was a remarkable person and to appreciate better *La Tectonique de l'Asie,* some discussion of his background is necessary. His main biographers—Maurice Lugeon, his teacher,[1] Charles Eugène Wegmann, his most famous student and his successor,[2] and P. L. Borel, a contemporary philosopher[3] all concur that to describe Argand is an almost impossible task. In fact, a genius is so multifaceted that many aspects of his personality are bound to escape the most penetrating analysis. However, I feel that by combining their three different approaches, it is possi-

* The address was published in the *Proceedings of the XIIIth International Geological Congress,* Vol. 1, Pt. 5 (1924), pp. 171–372.

xiii

ble to sketch a scientist whose personal life appears to have been utterly chaotic though he produced the most harmonious synthesis of the planet that geology has ever seen. In this introduction, I shall emphasize those details of Argand's personal life, of his career, and of his work that show the physical and intellectual conditions under which *La Tectonique de l'Asie* was created, the profound influence it had at the time of its publication, and the importance it gradually acquired in the evolution of ideas that led from continental drift to plate tectonics.

Argand lived during the glorious epoch in which the nappe structure of the Alps was being unraveled. He was well prepared, physically and intellectually, to play a fundamental role in the successful attempt to understand one of the most complicated mountain ranges of the world. Argand was a daring mountain climber, having since his youth engaged in that activity in the vicinity of Geneva, his hometown. Furthermore, he had received a broad education, studying architecture, medicine, and finally geology. During Argand's apprenticeship with an architect, his great skill as a draftsman became apparent. In fact, his grasp of spatial relationships and his memory of shapes were extraordinary. Throughout his career he would, with pen, pencil, brush, or chalk, sketch from memory maps (Figure 1), landscapes, and stereograms, as well as create portraits—some of them hilarious caricatures of friends and foes. His ability to think in three dimensions was to prove a major asset in the unraveling of the pattern of deformation through time of the most complicated shapes that Alpine overthrusts and recumbent folds could display. Despite Argand's talent for draftsmanship, his mother, a strong-willed and remarkable woman, decided against his being an architect or artist, and he began studying medicine, first in Paris, then in Geneva and Lausanne. In Lausanne in February 1905, after having read a few books on geology that he had come upon in the library of the Faculty of Medicine, Argand walked into the office of Professor Maurice Lugeon and declared his interest in Alpine geology. Of this first meeting, Lugeon, deeply impressed, wrote: "a man 25 years old talked to me with the science of a great master."[4]

Lugeon had just demonstrated the existence of overthrusts in the High Calcareous Alps (Helvetic nappes) and of recumbent folds in the Simplon massif. Although convinced that the Monte Rosa was part of a large recumbent fold, he had stopped short his investigations in front of the imposing massif of the Matterhorn—Dent Blanche. In these central regions, more than anywhere else in the Pennine zone, where metamorphism had obscured stratigraphy to a large extent, a structural approach was the only possible one. Here, the field geologist and the architect in Argand found their simultaneous challenge. Through intense mapping of these most inaccessible parts of the Alps, Argand solved the riddle that had stumped Lugeon: he demonstrated the existence of a Dent Blanche nappe

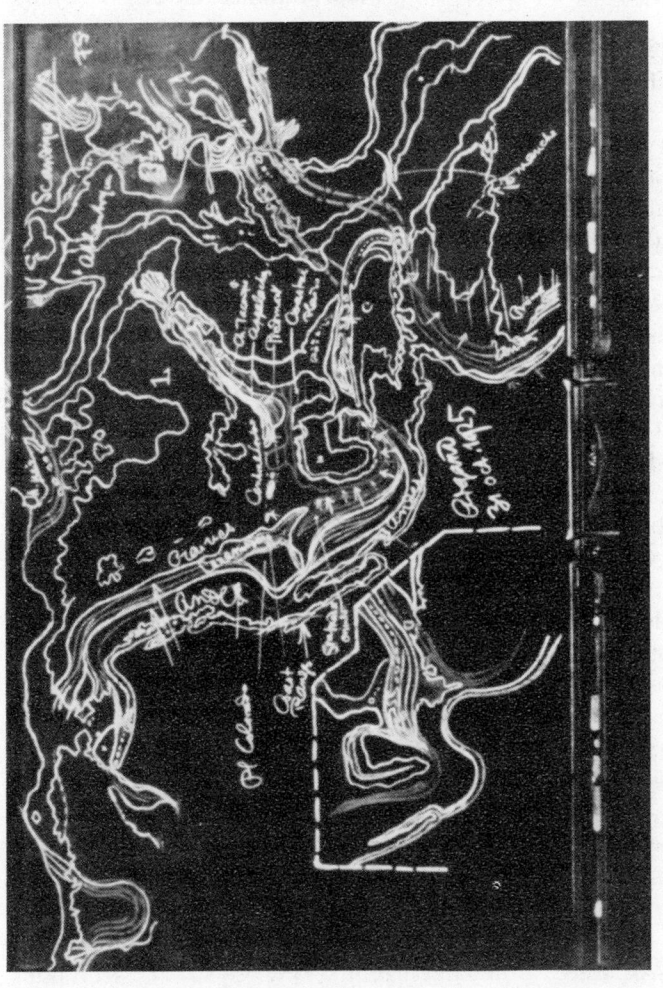

FIGURE 1. Blackboard sketch drawn by Emile Argand for his lectures on geotectonics, dated October 31, 1925. It shows the following structural features: the Canadian shield; the Taconic Range, representing the southern end of the Caledonian folding of Scandinavia, Scotland, and East Greenland; the Hercynian chain of Europe and the Appalachian chain of North America; and the Alpine system (the Rocky Mountains, the Laramides, the Coast Ranges, and the Antilles, as well as the Andes, Alps, and Mediterranean chains). Notice the structural connection between the Antillean and the Mediterranean chain cutting across the Mid-Atlantic Ridge. (Reproduced from M. Lugeon, *Bull. Soc. Neuch. Sc. Nat.*, 1940, Vol. 65, Plate II.)

overriding the Monte Rosa nappe. He achieved this feat by developing new geometric methods whereby to visualize the cylindrical shapes of recumbent folds deformed by axial culminations and depressions and by reconstructing them several miles above and below the present topographic surface. Thus he illustrated the "Pennine style," typical of the central part of many orogenic belts, in his maps and memoir of 1908 on the Dent Blanche.[5]

Having now developed a powerful technique of structural analysis, Argand expanded his field of investigation to the entire arc of the western Alps. In 1911, by means of four plates accompanied by a short commentary, he produced the graphic synthesis of that complex orogenic belt.[6] This was the year that Argand was appointed professor of geology at the University of Neuchâtel, succeeding Hans Schardt.

Argand's mind had been wandering beyond the western Alps since 1909, when he had seen in Lugeon's office a relief map of Eurasia. He decided to attempt a tectonic map of that immense and little-known area as a preliminary essay toward a tectonic synthesis on a planetary scale. Another factor in his decision was the strong influence exerted on him by two works of Eduard Suess, *Die Entstehung der Alpen* (1875)[7] and *Das Antlitz der Erde* (1883–1909);[8] the latter was an analytical description of the entire Earth, while the former, using the Alps as a model, set forth the principles of the interpretation of structural geology on a global scale.

Argand knew by heart *Das Antlitz der Erde,* having read the original edition in German. His memory of shapes, as already mentioned, was extraordinary; so was his language ability. He spoke six languages since his youth: French, German, English, Italian, Spanish, and modern Greek. Subsequently, he learned Russian, Sanskrit, an array of Slavic and Scandinavian languages, and eventually he mastered about 3,000 words of Chinese, some of which he learned to write. He also read Latin and ancient Greek, which enabled him to study classical works in the original. Indeed, Argand was a universal man in the style of Alexander von Humboldt.

The preparation of the tectonic map of Eurasia required from Argand the reading or at least the consultation of publications and maps by the thousand; it is estimated that he dealt with a total of seventeen languages in his compilation. When confronted with a paper written in a language he did not know, he would learn the language in a few days. Lugeon relates that Argand disappeared once for several days, during which time he apparently did not sleep. When Argand returned to the office he looked haggard but he had mastered Norwegian.[9]

A first version of the tectonic map of Eurasia was completed in 1912. Through the influence of Emmanuel de Margerie, who recognized very early the extraordinary qualities of Argand, the Spendiaroff Prize was awarded Argand in 1913 by the International Geological Congress meeting in Toronto. Although this map was never published, it is still in the possession of the Department of Geology of the University of Neuchâtel. Argand used it extensively in his lecturing on geotectonics.

Argand, not satisfied with a static description of Alpine or worldwide structures, began to add to his visualization the dimension of time. In a fundamental paper written in 1915, *Sur l'arc des Alpes occidentales,*[10] he developed a kinematic approach to orogeny that he called "embryotectonics." By means of sequential diagrams, he unfolded the recumbent folds, overthrusts, and other tectonic features back in time into the original cordilleras or other continental features from which they had arisen. He used the principle of conservation of the volume of the tectonic objects, and not the perimeters, since the unfolding of very stretched recumbent folds would lead to excessively large original surfaces. Furthermore, he assumed that the same tangential stress with increasing intensity culminating at the paroxysm directed the evolution from embryonic folds to the final complex recumbent folds. Thus Argand illustrated dynamic tectonics, presenting by means of a succession of images the continuum of structural deformation.

Although Argand's work of 1915 was replete with new structural concepts, these concepts were still considered to be within the leading geotectonic hypothesis of the time, namely, the contraction theory postulated by Elie de Beaumont and Suess. This was the "fixist" view, according to which the compression of mobile or geosynclinal zones between rigid continental masses and the related shortening of the crust expressed by nappes and overthrusts were due to a general decrease in the volume of the Earth.

The publication, also in 1915, of Wegener's fundamental paper containing his hypothesis of continental drifting[11] led to a complete change in Argand's thinking. It provided him with a new doctrine—"mobilism"—that satisfied to a much greater extent the space and time requirements of his synthesis than the concept of the general contraction of the Earth. Argand's carefully reasoned acceptance of Wegener's theory was also a bold political action. This was the 1916–1917 winter, the war was raging, and anti-German feelings were running high in Switzerland. In fact, it was strictly forbidden to read in public or private materials printed in Germany. Argand was the first in Switzerland to read Wegener and he presented Wegener's ideas as a surprise at the meeting on November 3, 1916, of the Neuchâtel Society of the Natural Sciences.[12]

Soon afterward, Argand began to prepare a second version of the tectonic map of Eurasia in the framework of mobilism in the Wegenerian sense (Figure 2).[13] He was again influenced by Emmanuel de Margerie, who apparently felt that at the end of the war, when international geological gatherings would resume, Argand should present his map, together with a major address, to a large audience. As the deadline of the Brussels

FIGURE 2. Emile Argand at work on the second version of his tectonic map of Eurasia, Neuchâtel, 1920. (Photograph by C. E. Wegmann, courtesy of J. P. Schaer.)

congress of 1922 approached, the pace of Argand's work on the second map became hectic. He would drive himself and his assistants for as long as two days without rest. His principal assistant, C. E. Wegmann, contributed extensively to the compilation of the new data and to the drafting of the map itself.

It is appropriate to point out that Argand obtained data from an unusual source—Swiss petroleum geologists who periodically returned home on vacation from faraway countries and paid him visits. At that time Swiss petroleum geologists were in great demand, and although Argand professed a slight disdain for this particular area of geology, it is clear that he profited from it. His visitors in turn did not leave empty-handed; a discussion with Argand of their observations always revealed interesting points that they had not previously considered.

During the last phase of preparation of the tectonic map, when the colored segments were glued on cloth, the colors smeared so badly that the document was practically ruined. In Brussels at the last minute, the map was fairly well restored by the Military Cartographic Institute of Belgium.[14]

It may be gathered from the circumstances of preparation of the map that Argand did not give much thought to the writing of the address; it is not known whether he made any notes for it. The actual writing of *La*

Tectonique de l'Asie, to be published in one of the volumes of the proceedings of the congress, turned into a long waiting game. Since no final text had been delivered after the oral presentation, as required by the regulations of the meeting, a steady flow of registered letters and telegrams poured into Neuchâtel as fretting officials in Brussels were holding up the printing of the volume that was scheduled to include Argand's address. Many friends intervened but to no avail. Argand, at the sight of the pile of letters and cables, would simply say, "I know what they contain," and without even opening them he would sweep them, with a theatrical gesture and much obvious satisfaction, into the wastebasket.

Without hesitation and as if by magic Argand could sketch the most complicated diagrams, but writing down his ideas was another matter. Having presented his thoughts in public, he could not bring himself to undertake the laborious task of putting them down on paper. Even his mentor, Maurice Lugeon, failed to persuade Argand to commit the address to writing. Indeed, the awareness that so many geologists throughout the world were eagerly expecting publication of the talk seemed to give Argand an exaggerated sense of importance, which he apparently enjoyed. An unexpected event finally provided a solution to this crisis of procrastination.

During 1923, Lugeon had to take a consulting trip and he asked Argand to teach his course in stratigraphy at the University of Lausanne. Argand accepted, but upon his arrival he was told that he would have to spend all his leisure time secluded in Lugeon's villa, "Les Préalpes"— with a well-stocked cellar, good food, and plenty of his favorite cigars— until completion of the text. His work would take place under the benevolent supervision of Mrs. Lugeon, whom Argand deeply respected. The stratagem worked, but the writing of *La Tectonique de l'Asie* was a harrowing experience. Argand attempted many times to burn in despair his newly written text but was stopped just in time by Mrs. Lugeon. Eventually the manuscript was completed. The cellar was supposedly half empty, but geology had acquired a unique document, the last page of which Argand presented to Mrs. Lugeon before returning to Neuchâtel (Figure 3). This was 1924 and the proceedings of the International Geological Congress could now go to press.

The title *La Tectonique de l'Asie* is somewhat misleading because the volume encompasses not only Asia but the entire world in a grandiose vision that attempts to relate all aspects of tectonic deformation in time and space, as if described by an ever-present observer who sees through millions of years a plastically deformed planet kneaded into mountain ranges of all types by the effects of continental drift.

The text itself is a delicate interweaving of abstract science and subtle poetry by a man who understood the most esoteric physical and mathematical aspects of mechanics, who responded to the beauty of lofty moun-

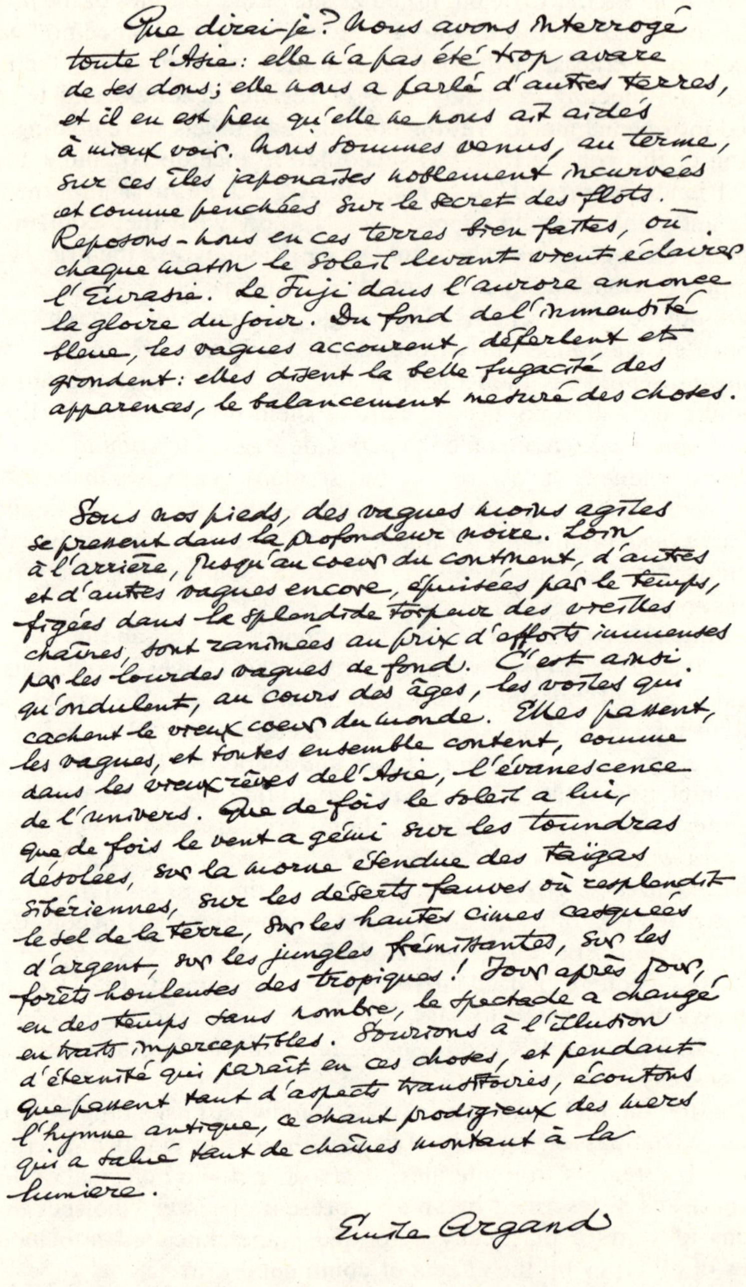

> Que dirai-je? Nous avons interrogé toute l'Asie: elle n'a pas été trop avare de ses dons; elle nous a parlé d'autres terres, et il en est peu qu'elle ne nous ait aidés à mieux voir. Nous sommes venus, au terme, sur ces îles japonaises noblement incurvées et comme penchées sur le secret des flots. Reposons-nous en ces terres bien faites, où chaque matin le soleil levant vient éclairer l'Eurasie. Le Fuji dans l'aurore annonce la gloire du jour. Du fond de l'immensité bleue, les vagues accourent, déferlent et grondent: elles disent la belle fugacité des apparences, le balancement mesuré des choses.
>
> Sous nos pieds, des vagues moins agitées se pressent dans la profondeur noire. Loin à l'arrière, jusqu'au cœur du continent, d'autres et d'autres vagues encore, épuisées par le temps, figées dans la splendide torpeur des vieilles chaînes, sont ranimées au prix d'efforts immenses par les lourdes vagues de fond. C'est ainsi qu'ondulent, au cours des âges, les vastes qui cachent le vieux cœur du monde. Elles passent, les vagues, et toutes ensemble content, comme dans les vieux rêves de l'Asie, l'évanescence de l'univers. Que de fois le soleil a lui, que de fois le vent a gémi sur les toundras désolées, sur la morne étendue des taïgas sibériennes, sur les déserts fauves où resplendit le sel de la terre, sur les hautes cimes casquées d'argent, sur les jungles frémissantes, sur les forêts houleuses des tropiques! Jour après jour, en des temps sans nombre, le spectacle a changé en traits imperceptibles. Souscrivons à l'illusion d'éternité qui paraît en ces choses, et pendant que passent tant d'aspects transitoires, écoutons l'hymne antique, ce chant prodigieux des mers qui a salué tant de chaînes montant à la lumière.
>
> *Emile Argand*

FIGURE 3. Facsimile reproduction of the last page of Emile Argand's manuscript of *La Tectonique de l'Asie*. (Reproduced from M. Lugeon, *Bull. Soc. Neuch. Sc. Nat.*, 1940, Vol. 65, plate intercalated between p. 48 and p. 49.)

tain peaks, and who could imagine the delicate festoons of tropical islands that he had never seen. The narrative does not exceed 150 pages, but a full understanding of it can be achieved only through close study of each sentence because extreme concision is Argand's rule. Words have precise meanings within carefully balanced sentences, and Argand sculptured every sentence to perfection. The difficulty of some of his passages arises from the very clearness with which a complex idea was present in his mind, and his strenuous effort to express the idea in its entirety. Argand's vocabulary is luxuriant, rich in terms freely borrowed from architecture, medicine, and the military arts. And when such terms appeared inadequate to express his three-dimensional picture of the world undergoing deformation, he did not hesitate to create new ones.

La Tectonique de l'Asie consists of three distinct parts: the main text, an extensive analytical table of contents, which is really an exhaustive index, and a set of twenty-seven illustrations accompanied by long explanations. The table of contents should be examined first since it is the thread that leads the reader through the intricacy of the text, in which definitions of fundamental concepts and actual examples are tightly interwoven. The main text should be considered a description of Argand's tectonic map (enclosed in a pocket at the end of this volume) and should be read in conjunction with a physiographic or recent geological map of Asia.[15] The set of illustrations is of fundamental importance because each picture is by itself so full of implications that its significance extends beyond its caption. These illustrations reveal to what extent Argand had mastered the pattern of structural deformation over large areas of our planet. His simultaneous use of a graphic and a written approach was not by any means redundant but had the purpose of insuring that all his new tectonic concepts would be fully understood by the reader. Figure 19C, a cross-section of the Alps, is the key to his work; from it were derived Figures 19A, 19B, and 17, which show the relationships among the Apennines, the Alps, and the Atlas Mountains, and finally Figure 13, which deals with the Himalayas.

An analysis of *La Tectonique de l'Asie* can only touch upon a few major aspects; because of the concise nature of the presentation a complete appraisal would be a virtual restatement of the text. Argand's contribution rests on two fundamental concepts of Wegener's—first, the hypothesis of continental rafts of sial floating and drifting on the sima; second, the concept of the plastic behavior of all crustal materials under long-continued stresses. Because for Argand deformation originates from continental drift, it can consist only of horizontal stresses; hence, the duality of epeirogenic and orogenic movements is an illusion, while radial movements are but local and minor components of the horizontal effort.

There are four major ways by means of which the plastic continental rafts develop zones of folding. Basement folds *(plis de fond)* express the large-scale doming of continents through friction at the base of the drifting crust. Basement folds require the greatest consumption of energy, and the large volume of reactivated old material they involve makes them the most important form of deformation of the crust. Marginal folds *(chaînes marginales)* are generated at the bow of drifting continents, where the less plastic mass of the continent is folded by the almost paradoxical "resistance" of the more plastic substratum. According to a new hypothesis, geosynclinal folds *(chaînes géosynclinales)* were interpreted by Argand to be the result of plastic behavior under tension followed in time by compression. Initial subsidence in the mobile belts implies thinning of the sial by tension; when compression occurs, the thin, weakened floor of the geosynclines, together with the freshly deposited sediments, are changed into elegant recumbent folds. Finally, the sedimentary covers of cratonic masses fold disharmonically over their substratum by decollement generating cover folding *(plissements de couverture)*.

In conclusion, Argand explained the entire tectonic deformation of the globe in terms of a systematic application of differences of relative plasticity under the action of continental drift. Although his major interest was the analysis of deformations belonging to the Hercynian and Alpine cycles, Argand attempted to unravel older episodes of drift involving the Precambrian shields of Africa and Eurasia but with much less success because of insufficient data on these older rocks, particularly in regard to geochronology.

In retrospect, one can perceive several flaws in Argand's concept of global tectonics. One is the exaggerated degree of plasticity he attributed to the continental masses, another to have drifted continental masses to an extent not warranted by the data available at that time. But the most critical flaw is that he considered the style of deformation as essentially the same from the deepest to the shallowest parts of the bodies undergoing deformation. Argand's most brilliant student, C. E. Wegmann, demonstrated in association with the school of Scandinavian geologists, mainly through the investigation of Precambrian shields, that the style of tectonic deformation changed vertically in a significant manner, leading to the concepts of infrastructure, tectonic levels *(étages tectoniques)*, etc. Interestingly, Argand visited Finland in 1931—on one of the very few trips he ever took—invited as honorary chairman of the congress on the study of the Precambrian and of the tectonics of old platforms. Thus, he was able to see what Wegmann had to show. After a week in the field and despite his initial surprise, Argand returned home seemingly convinced of the vertical variation of tectonic style and genuinely delighted by the contribution made by one of his students. In reality, the geology of Precambrian shields remained foreign to him; his picture of the world's tectonics was and remained Mediterranean.

Upon publication of *La Tectonique de l'Asie*, some skeptics could not accept the fact that a man who had traveled very little beyond the Mediterranean could interpret the tectonics of such remote areas as Siberia and Mongolia. Gradually this incredulity disappeared as demonstrated, for instance, by a translation into Russian in 1935 of Argand's work,[16] which shows that the geologists of that part of the world considered his contribution fundamental to understanding the geology of Eurasia. Moreover, the impact of Argand's work on European geological thinking was so profound that his approach and his terminology, together with Wegener's theory, became a basic part of geological education in most universities. As examples of this impact, it is sufficient to note *The Structure of the Alps* by L. W. Collet (1927, 1935)[17] and *Our Wandering Continents* by A. L. Du Toit (1937).[18] Only North American geologists remained fixists, with a few exceptions such as Walter H. Bucher, and paid little attention to Argand's views (as well as to Wegener's). In Bucher's volume, *The Deformation of the Earth's Crust* (1933),[19] Argand's contribution, although not fully accepted, was presented as the work of a genius and given an extensive presentation accompanied by the reproduction of several important illustrations.

After publication of his masterpiece, Argand wrote very little, mainly a few short notes and explanatory comments on quadrangle maps of the Alps. The reasons for this reticence are fairly complex. Unquestionably, a heavy teaching load which included mineralogy, and, apparently, the pettiness of a small university community, which could not easily accept the many sides of his unusual personality, played an important role in Argand's retreat. Paradoxically, the most important factor was his own success, which led many European geologists to borrow heavily from his new concepts without giving Argand the credit he deserved.

While Argand withdrew from the mainstream of geology, his interests turned toward advanced mathematics, physics, philosophy, and the comparative study of languages. Sometimes his absorption in, and fascination with, a new subject interfered with his teaching activities: he once disappeared for four weeks to return only to tell his assistant Wegmann that he had bought the translation of the *Arabian Nights,* that he was leaving for another three weeks in order to complete the reading of the second half of the volume, and that Wegmann should see to it that Argand's students were kept busy. One can see in the choice of the mature Argand's studies a renewal of some of his youthful interests. Still, he was deeply concerned about finding a rational explanation for continental drift. Why did the continents move? What was the driving process and the origin of the required energy? Unfortunately, he died in 1940 without leaving any documents to indicate how far he had advanced in solving this fundamental problem, which he wanted to approach alone, away from interference from the outside world.

In the evolution of geological concepts from continental drift to plate tectonics, *La Tectonique de l'Asie* is a milestone.[20] In the fifties, it was often quoted but probably little understood. At present it is ignored, even in the study of Alpine orogeny in the Mediterranean, where Argand postulated, as the basic mechanism, a compression between Africa and Europe closing the Tethys. He appreciated the complexity of the Mediterranean tectonic puzzle, as is clearly shown in his maps (see Figures 20 to 27), and accordingly introduced in his reconstructions a number of compressional, extensional, and rotational situations to explain the observed facts. Obviously, as pointed out by J. F. Dewey and his collaborators (1973),[21] at the time Argand was working, the nature of the drift in the Atlantic was not known; therefore, the time and space pattern of the relative movements between Africa and Europe could not be accurately established. Nevertheless, some of the mechanical artifices created by Argand bear interesting similarities to those in use today, devices that are based on the idea that the Mediterranean consists of a complex assemblage of microplates that, although undergoing a relative northward compression from Africa, lead, as an effect of their differential motion, to a complex reaction along the southern margin of Europe. As indicated by Dewey and his associates:[22] "There probably never was only a single plate boundary between Africa and Europe; but rather, there was at all times a network of compressional, extensional, and transform boundaries." In the context of his time, Argand came close to such modern views.

Argand also reached far beyond his time when he wrote in Section XXVII of *La Tectonique de l'Asie* that he had undertaken the confrontation of fixism and mobilism for the entire Earth, that is, on a global scale. He forecasted plate tectonics when he wrote in the same section that other mobilistic theories would be born or reborn in rejuvenated forms in the future. Therefore, Argand should be credited at least in part with the following statement made in 1970 by J. F. Dewey and J. M. Bird:[23] "The emergence of the theory of lithosphere plate tectonics, the new global tectonics, for the first time provides a unifying worldwide explanation for tectonic processes."

The resurrection of the numerous modern aspects of Argand's global tectonics is long overdue, as pointed out recently by Marcel Roubault.[24] I offer this translation to the English-speaking geological community to enhance the appreciation of a man who was important in the shaping of our modern concepts of structural geology.

Notes

1. M. Lugeon, "Emile Argand," *Bull. Soc. Neuch. Sc. Nat.*, Vol. 65 (1940), pp. 25–53.

2. C. E. Wegmann, "Emile Argand," *Dictionary of Scientific Biography* (New York: Charles Scribner's Sons, 1970), Vol. 1, pp. 235–237.
3. P. L. Borel, *La Symphonie intérieure,* Vol. III, *Orgies philosophiques* (Neuchâtel: Editions H. Messeiller, 1974), pp. 131–166.
4. M. Lugeon, op. cit., p. 26.
5. E. Argand, "Carte géologique du massif de la Dent Blanche," *Mat. carte géol. Suisse,* N.S., Livr. 23 (Berne, 1908) (carte spéciale No. 52).
6. E. Argand, "Les Nappes de recouvrement des Alpes occidentales et les territoires environnants—Essai de carte structurale 1 : 500.000," *Mat. carte géol. Suisse,* N.S., Livr. 27 (Berne, 1911) (carte spéciale No. 64).
7. E. Suess, *Die Entstehung der Alpen* (Vienna: W. Braumüller, 1875).
8. E. Suess, *Das Antlitz der Erde,* 3 vols. in 4 (Prague: F. Tempsky, 1833–1909).
9. M. Lugeon, op. cit., p. 28.
10. E. Argand, "Sur l'arc des Alpes occidentales," *Eclogae Geol. Helv.,* Vol. 14, No. 1 (1916), pp. 145–191.
11. A. Wegener, *Die Entstehung der Kontinente und Ozeane* (Braunschweig: F. Vieweg und Sohn, 1915) Sammlung Vieweg No. 23.
12. Meeting of November 3, 1916. Argand presented the modern views on the origin of the continents and the oceans. He discussed the theory of Wegener, which attempted to replace the theory of collapses by that of dislocations. This new concept explains much better than the older one the similarities of fauna and flora between regions today separated by extensive oceans. It is similarly supported by the study of the late Paleozoic glaciations. See *Bull. Soc. Neuch. Sc. Nat.,* Vol. 42 (1916–17), p. 115.
13. E. Argand, *Carte tectonique de l'Eurasie* (1922), published by the XIIIth International Geological Congress, Brussels, under the supervision of Armand Renier, general secretary. Color photographic reduction from the original scale of 1 : 8,000,000 to 1 : 25,000,000 (Brussels: Guy Onkelinx, 1928).
14. See Section II of this volume.
15. *Tectonic Map of China and Mongolia,* compiled principally by M. J. Terman, 2 sheets, 1 : 500,000 (Boulder, Colo.: Geological Society of America, 1973).
16. E. Argand, *Tektonika Azii,* Doklad na Brîussel'skoĭ (XIII) sessii Mezhdunarodnogo geologickeskogo kongressa v. 1922 g [Tectonics of Asia, Report of the XIIIth session of the International Geological Congress, Brussels, 1922], translated by I. P. Teplîakovaia, under the editorship of ÎU. M. Sheinmann (Moscow and Leningrad: Ob" edinnoe nauchno-tekhnicheskoe izd-vo NKTP SSSR, general editorship of the Geol. i geodezicheskoĭ lit-ry, 1935).
17. L. W. Collet, *The Structure of the Alps,* 2d ed. (London: E. Arnold and Co., 1935; first pub. 1927; reprint of the 1935 ed. with introduction by A. V. Carozzi, Huntington, N.Y.: Robert E. Krieger Publishing Co., 1974).
18. A. L. Du Toit, *Our Wandering Continents* (Edinburgh and London: Oliver and Boyd, 1937).
19. W. H. Bucher, *The Deformation of the Earth's Crust* (Princeton: Princeton University Press, 1933; reprinted., New York: Hafner Publishing Co., 1957).

20. A. Hallam, *A Revolution in the Earth Sciences, from Continental Drift to Plate Tectonics* (Oxford: Clarendon Press, 1973).
21. J. F. Dewey, W. C. Pitmann III, W. B. F. Ryan, and J. Bonnin, "Plate Tectonics and the Evolution of the Alpine System," *Bull. Geol. Soc. Am.*, Vol. 84, No. 10 (1973), pp. 3137–3180.
22. Ibid., p. 3139.
23. J. F. Dewey and J. M. Bird, "Mountain Belts and the New Global Tectonics," *Jour. Geoph. Research,* Vol. 75, No. 14 (1970), pp. 2625–2647, see p. 2626.
24. M. Roubault, *La dérive des continents* (Paris: Presses Universitaires de France, 1972; collection "Que sais-je?" No. 1503), pp. 114–124.

TECTONICS
OF ASIA

Introductory Comments: Orogenic Cycles

Gentlemen, twelve years have elapsed since the writing of the last pages of the great work by F. E. Suess,[1] twelve years during which numerous distinguished works on a great variety of regions of the Earth have been published, works that provide a priceless amount of new facts and occasionally bold interpretations. If the master were to return among us, he might perhaps be inclined to retouch a few details of the plan, to perfect a given sketch, or to account for a given episode by means of a different presentation. However, I assume that he would not consider changing the major lines of the monument because its arrangement has, to a great extent, remained correct.

Marcel Bertrand, whose views embody depth and force, has talked about Caledonian, Hercynian, and Alpine foldings.[2] This order and these subdivisions, which display the sureness of a master, shall remain. They apply to the entire Earth. Setting aside for the time being the Precambrian foldings, which I know have been active through numerous repeated cycles, I shall mention, as a matter of introduction, in what sense Caledonian, Hercynian, and Alpine foldings shall be considered here. In each one of the three cases, I visualize a complete orogenic cycle with its long preparation consisting of precursor foldings, its paroxysms with folds reaching their maximum development, its decline marked by late folding phases. Whenever in regard to these spans of time I shall wish to stress a major phase or one less prominent, whenever I shall point out some brief episode intercalated among them, I shall say so. Besides, it is inconceivable that the limits of a cycle could be absolutely clean-cut and everywhere perfectly contemporaneous. In fact, excellent reasons exist for us to consider just the opposite situation. Therefore, let me retain in these matters the option of establishing time limits that, as indicated by the facts, shall always be narrow. These details set aside, the Caledonian foldings result from an Early Paleozoic cycle; the Hercynian foldings from a Late Paleozoic cycle; the Alpine foldings from a cycle that spans Mesozoic, Cenozoic, and Quaternary times, including the Present.

The opportunity to reinterpret on a large scale and in detail the large continental masses arises from the progress made in these recent years in establishing the time succession of facts and also—there is no exaggeration in saying so—in the art (172)[3] of their interpretation.* Therefore, a few moments shall first be devoted to this particular point.

---

* For reference purposes, the page numbers of Argand's original text are retained in the English translation and are enclosed in parentheses. Each page number refers to the material following it.

I. Art of Interpreting Structural Facts

The volumes, the surfaces, the lines—in one word, the structures that build a tectonic construction—do not represent the whole picture: there is also the movement that animated and still animates these bodies because the history continues and we live under no particular privileged conditions at any given time in this great process. One might say that there is a static tectonics and a dynamic tectonics. The first is the art of defining the present state of the structures; this merely requires correct observations, complemented—through means I shall not discuss here—by adequate comparisons and interpretations. This kind of tectonics, because of its static attitude toward the world, is not self-sufficient since the world does not stop. The dynamic tectonics would be, in its final expression, a completed tectonics, that is, a continuous history of the deformations of the planet in which all the evidences would relate to each other without gaps. This ideal situation is unachievable. Nevertheless, one will still strive toward it in the future if static tectonics, as a necessary starting point, is explained as precisely as possible and if the art of restoring movement to it, an art that is among the most delicate and subtle, is applied correctly. Because you cannot see movement as you see structures as objects standing in front of you, you have to recreate this movement in your mind and guide it in such a way as to explain, through time, the preserved evidences, and finally suggest it by means of an image.

This presentation shall deal more with real tectonics than with orogenic theories, the latter being really attempts at molding into too rigid formulas a matter that is too polymorphic, thus reducing to simple physical laws the unlimited and so little known mobility of structural objects.

Naturally, this is an effort to understand, although in an entirely different sense from that implied by the original force of the word "theory." My ambition, which is more realistic, is to reinterpret in a more precise fashion static tectonics and to reveal dynamic tectonics. I do not pretend to reduce tectonics to physics—this is a matter for the future. I wish only to suggest, insofar as the still enormous gaps in the evidences permit, the picture of shapes in movement, as well as the vision of an unfortunately incomplete history.

This history, visualized as a whole, would consist entirely of tectonic deformations taking place in the three dimensions of a space always devoid of gaps through geologic time. These tectonic deformations would relate to each other during an instant of time in a perfectly clear set of interactions, and from one instant of time to another, in a fully developed determinism; they (173) would relate to each other all the events of a given orogenic cycle, and each cycle to the preceding one; they would regulate, at any time, the general features pertaining to the substratum and the nature of both continental and marine sediments and to the geomorphologic condition of the areas; they would account for and explain the stratigraphic succession and the geomorphologic organization. This movement of images, comparable to treetop flying above the facts, would be like the key of the history; it would take hold of the data and gather in its flexible order all their far-reaching implications; it would generate the explanation of things not by means of a precise theory but as a true story. Therefore, one would see all the phenomena with all their relationships as an ideal and privileged spectator who would have watched the unfolding of this history and would have been able to condense it into a more rapid succession of images. Let us accept this new aspect of tectonics entirely subjected to facts and yet so easy to grasp, as clearly shown by a limited but very complicated object, namely, the recently reconstructed history of the movements in the western Alps.[4] I say "accept this new aspect" and not reach a final stage; the latter shall be approched only gradually, but this is the way to proceed. Thus are outlined, far above our present knowledge, the ideal goal of so many efforts and—for the time being—the approximate direction.

Such broad approaches become fruitful only if one analyzes harmoniously the web of facts on all the scales of tectonics: continental scale, scale of the chain, of the nappe, of the fold, even of the thin section. Furthermore, such approaches are valid only if supported first by views of variable significance that belong to classical geology and are largely accepted by myself, and second by those I have created, renewed, or simply revised through personal experience. Therefore, I am inclined in these initial statements to mention briefly some of the latter views, thus appreciably simplifying the presentation of the regional interpretations that follow. I beg to be forgiven in this short presentation for the apparently absolute position suggested by such concise statements: this is not the real nature of my thinking, and the nuances that are inseparable from such a mobile and delicate order will be expressed at the appropriate place, and for each object in particular, as I have attempted to do elsewhere.

First, restitutions of structures and movements that are limited to one or two dimensions always remain more or less analytical or episodic. There is no tectonic synthesis without the vision of a continuum in three dimensions undergoing deformation.

Second, mountain ranges cannot all be equated to a single type of structure; it is higher up, within the movement itself, that the unifying and explanatory principle lies. Not all chains (174) arise from geosynclines: those that originate from them usually show in their initial stages an interplay of furrows and cordilleras, more or less arcuated and of island-arc type; their orogenic paroxysms represent the exaggeration of such movements and of such forms, often carried as far as piled-up nappes; their late foldings correspond to the moderate and often repeated accentuation, in the style of previously acquired deformations. Precursor foldings, paroxysmal foldings, and late replicas all make up, through geologic time, an orogenic cycle. And the history of a mountain range, whether geosynclinal or not, is, when the entire cycle is considered, but the continuously renewed deformation of its folds. At any instant of time, this behavior, by means of one of the most delicate effects, regulates the horizontal distribution of the deposits that form within the chain and around it at a short or a long distance; at each instant, the deformation also varies, and its regulating function therefore extends to the vertical distribution of the sediments and of their changes of facies. This applies also to the physiographic condition of the emerged areas: backbones of embryonic cordilleras or of almost completed chains; sections of forelands warping under the horizontal push transmitted by the chain or bending in massive basement folds. Since the obviously tangential stress that is responsible for all these features displays during an orogenic cycle a great number of maxima and minima, of renewals and weakenings, one can see that the physiographic evolution of these objects will include as many geomorphologic cycles as episodes of renewal; that these cycles will be very unequally developed; and that each great subdivision of an orogenic cycle—embryonic period, paroxysms, period of replicas—will in general include numerous geomorphologic cycles, without implying—at least until this point of my talk—any pure vertical movement, whether epeirogenic or not. Indeed, the tectonic objects, while shrinking under tangential stress, are compelled to rise or to sink while warping. Therefore, a vertical effect is directly derived from the tangential stress, and too frequently this effect has been interpreted as a vertical movement independent of the deformation. During times of weakening of the tangential stress, the tectonic objects, laterally less supported, preserve at depth the acquired foldings. They display, throughout their entire mass and with the help of isostasy, small and more or less vertical movements that in fact result from tangential actions but give the misleading impression of pure uplifting or sinking. If one investigates the problem thoroughly, the reason for these illusions appears quite easily. The vertical component of the deformation, very apparent in its geomorphic or stratigraphic effects, is often much more obvious than the deformation in the two horizontal dimensions, particularly in the case of small precursor or late movements: (175) of the three contemporaneous effects of the same deformation, two are neglected and most commonly remain undetected; the third one, mentally isolated, appears as an independent or even late phenomenon. But nature, which operates only in volume, does not pay any attention to these analytical artifices. Being able to put back the *disjecta membra* of the analysis into the moving flux of things, and the discontinuous detail into the continuous whole, that is the secret: if one does not think along such lines, it may happen that regional tectonics, in reality harmoniously molded in one episode, will be presented as a chaos of epeirogenic movements and radial faults.

Third, I may say that the same illusions are not uncommon in the treatment of another problem that encompasses, together with the origin of the axial inclinations of folds, the mode of emergence of chains. It is not only in ground plan that chains, nappes, and folds undergo the influence of obstacles: it is, in fact, in volume. The plastic flux, while flowing, backs up behind obstacles: therefore, it will—all other things being equal—rise higher in the segments with obstacle than in those in which flowing remains less hindered. In the first case, it will generate, except under particular circumstances, an axial doming; in the second case, a less exalted segment. In general, longitudinal cross-sections do not display an unequal raising of chains, nappes, or folds that would have occurred after folding in the manner of an epeirogenic movement: this raising and lowering of axes are only the vertical aspects of the folding itself and are naturally contemporaneous with it.

Fourth, there may be danger in exaggerating the significance of vertical faults and in overestimating their importance with respect to the large deformed structures within which they occur. Unquestionably, innumerable fractures, more or less vertical and displaying variable throws, skirt or intersect horsts, grabens, domes, and basins of any radius of curvature and dislocate tabular coverings. Later all these features shall be referred to appropriately, but now I would like to look at them from a particular viewpoint. If one decides to reconstruct all these objects on their true scale, without exaggerating the vertical dimension, the faults with the greatest throw become minute details within immense structures folded on a large scale. One can already perceive that the real process is the deformation in volume and not the fracturing and also that the latter requires much less energy than the former and actually could be considered as one of its minor effects. This impression seems to be greatly reinforced if one can believe that warping and fracturing are contemporaneous, and there is, as I shall demonstrate later, more than one way to shed light on these different points. What becomes of a fracture with great throw along a dome such as the Baltic shield; or even along the margin of a smaller but very strongly warped block such as the Colorado Plateau; (176) along the margin of an Andean chain of great tonnage such as the Sierra Nevada; or finally along the small voussoirs such as the chainlets of Gobi or of the Great Basin? It is, therefore, conceivable that in spite of these examples, to be followed by many others, radial dislocations occur in the world; yet, as surprising as this might sound, I must add that nothing is known about them. I may say in anticipation that in the present state of science, the interpretation of vertical faults by radial stresses relies only on the impossibility of proving the contrary; whereas their explanation by means of tangential stresses, in which folding represents the original process, can be reached through positive criteria. For the time being, I shall leave out the great vertical faults that are supposed to be of distensible nature, namely, those explained by pure traction in a horizontal sense, that is, by tangential traction.

Fifth, it is useless to argue in favor of pure vertical movements in the case of tabular lands because strictly speaking they do not exist. Those usually called so are very gently folded lands, but this situation does not mean that the required energy was small: in fact, the energy involved in these deformations with large radius of curvature can be of considerable magnitude only because the deformed volume is large. It is impossible to visualize within old areas absolutely rigid massifs, and similarly within *living chains* entirely plastic flows. The concept of tabular lands is only a theoretical limit, useful for the gross expression of certain aspects—but nature does not take it upon itself to realize such a concept in its perfect purity. There is no tabular land or any other land that has not been compressed in the horizontal sense. Whether such a compression generates narrow folds within a well-ordered chain, or basement folds independent of ancient dead folds within an ancient infrastructure, or folds with any radius of curvature in the coverings of these old frames, one still deals with folds and with orogeny. Therefore, should one say that epeirogenic movements, apparently eliminated from living chains and even from epicontinental areas, should be disposed of? Can these movements remain inconceivable as the result of readjustments due to isostasy or to changes of volume in underlying magmas, beneath young chains as well as old basements? Can their vertical effects add themselves to those related to folds, either at the same time or after folding? I shall answer these questions as well as possible after a realistic interpretation of the deformation of continents. We shall see that there are no vertical movements that cannot be considered as the direct or indirect effects of volume deformations in which horizontal stresses are usually predominant. Meanwhile, one can see that the diagnosis of orogenic effects, within old basements as well as within living chains, can be made through positive geologic evidences, (177) whereas the idea of originally vertical movements can be sustained only in the absence of such evidences. One can immediately perceive in what manner folding is a general function and how its importance grows in size and in efficiency: it encompasses not only the living chains and the important reworking of previously folded old frames but also, in these two types of lands, the ordering of stratigraphy and of geomorphologic cycles. In conclusion, it seems reasonable to speak only with extreme caution about original vertical movements and particularly about epeirogenic movements. In problematic cases, it would be wise to talk only of deformations with large radius of curvature. One must agree that it would be difficult to exaggerate our ignorance of the physical causes of terrestrial deformations. But the radius of curvature, regardless of its size, is a fact that can be perfectly grasped, in a static as well as in a dynamic fashion.

I shall make a sixth point: the island arcs, the island festoons, and the ordered chains that are displayed by the present face of the Earth have reached very variable degrees of development or of reworking. These moving objects have the admirable property of showing side by side, at this instant in time, all the transitional aspects that one of them can display during its own history. What a spectacular example of comparative anatomy and what an unusual museum of tectonic embryology! One perceives great lessons to be drawn from such an important fact.

Seventh, the phenomena described under the name of virgation have acquired in recent years a new interest: they have revealed themselves particularly suitable for the diagnosis of the most delicate behaviors that an arc or a row of arcs can display in ground plan, and I shall have many opportunities to discuss this subject later.

Eighth, one has often thought of recognizing in several chains transverse folds of variable intensity. But the generation of transverse folds within an ordered chain is mechanically inconceivable except in the following, recently discovered case: in an arc of a certain width that is compelled to accentuate the curvature of its plan. The external side, undergoing extension, stretches while the internal side, submitted to longitudinal compression, may display small-scale transverse folds that are obviously related to the behavior of the main object and do not constitute an independent episode.

Ninth, one has often interpreted as transverse folds the trace, necessarily transversal in certain regions, of nappes that are perfectly ordered along the longitudinal trend of the chain. The transversal orientation of the traces—and not of the folds—results in these cases from the intersections between the nappes and the topographic surface, and the axial dipping plays its role there. These circumstances, when correctly interpreted, not only lead to the discovery of nappes in areas considered autochthonous but also eliminate even the appearance of transverse folds. (178) Obviously, in poorly known lands, or in the geology of islands on which so many features remain hidden as if devoid of relationship, the existence of transversal traces, when unexplainable by the occurrence of previously folded fragments incorporated in the chain, or by regional curvatures of the plan, or by the very unusual effect of true transverse folds, shall indicate overthrust nappes, whose existence it is always good to discover. Furthermore, it will lead, in the case in which axial dipping is favorable, to the identification of very thick pilings up of nappes, as happened in the western Alps, where it is very easy to draw, with a reasonable approximation, archbends buried at a depth of twenty kilometers.[5]

Tenth, backward foldings, which for such a long time have seemed an enigmatic exception to the unilateral overfolding of chains, have been related to this latter mechanism of which they represent only a peculiar case that usually originates from the reciprocal interference of folds and nappes, from the fact that this interference occurs in periodical phases, and particularly from the downward displacement, during the history of a chain, of the point of application of maximum tangential stress. Thus, one gets rid of the complication that would consist of considering, for the same chain and in the same homogeneous stress field, two distinct and opposed thrusts. Reconsidering now the regions poorly known as well as the tectonics of islands, one can visualize the great caution required in identifying the direction of overfolding: with only scattered observations, one shall not always know, lacking any other critical appraisal of the evidences, if a given point for which the direction of overfolding is known belongs to the predominant direct regime or to the subordinate retrograde regime. Any modern speculation on the direction of overfolding of the arcs may therefore become unreliable even when a few scattered data are available. This is particularly true when the overfolding is inferred only from the general shape of the arcs. In my opinion, for poorly known areas it is only when certain forms of virgation are generated that the general direction of the thrusting may be inferred, even in the absence of archbends.

Eleventh, a few words should be said about cases in which the limit between two successive orogenic cycles is not marked, according to the general rule, by a directly visible angular unconformity. This happens in places in which the folds of the first cycle have not been pushed up to emergence or in which the emergence has not been carried far enough for the erosion to reach—beneath the carapaces and the crests consisting of horizontal beds—the more tightly folded structures. This does not preclude the fact that the foldings of this first cycle might have been very intense: the point is that it is often possible, in spite of the absence of the great common criterion, to identify these foldings. According to circumstances, they will be displayed at depth by the conditions of positioning of the granites of (179) the first cycle and of the extent of their aureoles; at the surface, by a penetrating analysis of the horizontal and the vertical distribution of the sediments, by well-conceived unfoldings that will lead, for a given period, to the restitution of dissymetric embryonic cordilleras, by stratigraphic gaps restricted to geanticlines and expressed by either a disconformity or very weak angular unconformities made visible by transgressive or regressive wedges, with or without conglomerates, and more exceptionally—and only on top of the most eroded geanticlines—by true angular unconformities, always of limited extent, whose areas are distributed longitudinally. All these features—except the last one, which nevertheless remains perfectly conceivable—have been observed in the Pennine zone of the western Alps, and ideas derived from them are certainly going to be useful elsewhere. One deals here with a geosyncline that at the end of a first cycle—the Hercynian cycle—is almost filled by deposits and by folds, with the emergence of marginal areas and of geanticlinal crests with often horizontal beds, while at depth the folds are much more accentuated: one more episode and the geosyncline will sink anew, and its folds will become exaggerated during the Alpine cycle, without much discontinuity between past and future. In a geosyncline of such long duration, the conformity of deposits will naturally be the rule during more than one cycle; in the deep orogenic furrows, this will be an almost absolute law. In a more general way, one should search in any geosyncline that has succeeded in maintaining itself through more than one orogenic phase or more than one orogenic cycle—the Himalayan geosyncline, for instance—for often tenuous traces by means of which passed phases and cycles express themselves to us. Archbends and disconformities shall always be the most obvious evidences of folding, but whenever these are absent there are substitutive features whose fine resolution and power should not be neglected.

As my twelfth and last point, for all the recorders of the pulsations of folding such as archbends, unconformities, warpings of peneplains, and horizontal and vertical variations of marine and continental sediments, I would like to raise the question of their sensitivity to deformation. In this matter, I shall let speak the regional or local evidences, pointing out, nevertheless, the great variations that marine deposits can display in this respect. Assume that two folded wrinkles identical in all respects except their bathymetric condition undergo tangential stresses with the same vertical effect: this effect will be recorded by numerous variations among the neritic or littoral sediments if the crest of the wrinkle is close to sea level; if the crest remains at a certain depth, the recording will be either very weakly marked or even entirely absent. The portion of the foldings that operates near sea level is revealed in a somewhat enlarged fashion; whereas the deeper portion appears in a more subdued manner. All (180) these aspects should be carefully compared with each other if one wishes to grasp the movements in their true proportions. In order to reach such a goal, all of the criteria that I have just assembled in a delicate bundle appear necessary.

These are, in a very concise summary, some of the viewpoints that guided me in this investigation. I believe that a science aiming at a correct interpretation and a correct coordination of the numerous local geologic data should never afford to neglect any of these viewpoints. As I have tried to show, these variably encompassing viewpoints should all tend to fit into a broad vision that would be in an ultimate approach to tectonics the visualization of the total deforming movement.

The art of resurrecting, if only as an image, the complex interplay of the deformations reveals, as I have said previously, the animating factor of tectonics as a whole: from this re-created center may be perceived also, as within a moving framework in which everything develops without rest, the dynamic and tectonic conditions of the stratigraphic history and of the geomorphologic development; the same applies, up to a certain point, to the conditions of positioning of the magmatic rocks, at depth and at the surface. To rise to such a viewpoint, and from there to descend to the known facts of all orders of magnitude, and to explain them—this is almost all of geology and the essence of my thinking.

Does this approach have any chance of encompassing the entire Earth? Will this path lead to successes whose boundaries would be but those of the planet? I have no doubts in that respect: in this great interplay of tectonics, stratigraphy, and geomorphology, the leading agent is dynamic tectonics, namely deformation.

To encompass by means of such viewpoints that which is known about the major aspects of our Earth is still within the reach of a single man. Is it necessary to stress that in such a case one owes to others all the subject matter that one merely tries to mold into a new shape? Is it also necessary to point out the incomplete, the relative, and the temporary nature of such an attempt? Incomplete because of huge gaps that shall never be entirely filled and that interrupt the fabric of human knowledge: even these gaps have their value because it is often in such places that problems arise. Relative, because the quality of such an attempt depends on that of local geological data. Temporary, because science is being enriched every minute by new facts and because at any given time in this enrichment the degree of approximation, variable from one place to another, cannot equate for the whole what has been reached in a given limited area. Nevertheless, if the approach is good, so will the attempt be: the story of this embracement with its strength and its imperfection remains possible. This story will not be told today. At this time it is out of the question to say everything. It will be more realistic to sketch lightly, dealing mainly with one of our greater continental masses, (181) some of the lineaments that allow me to write this account. In spite of long experience with this kind of investigation, I would have certainly delayed the completion of such a task if the organizing committee of this congress, by its distinguished show of interest in these problems, had not honored me by this invitation.

II. Tectonic Map of Eurasia

The poster in front of you is the original of the tectonic map of Eurasia on a scale of 1:8,000,000. Two drafts were prepared—the first in 1912; the second, in 1922, was required by the rapid progress in geological exploration. Of these two successive approximations, it is naturally the later one that is presented to you. Needless to say, I have tried in both of them to neglect none of the available sources; the number of those that I was unable to consult is very small. Without giving a technical presentation, which would better fit an explanatory text should the map be published, I shall describe here the essentials of the aims pursued and of the graphic symbols used: to reach an appreciation of regional tectonics; to outline their most important features; to correlate them cautiously; to derive from such correlations general views and from the latter whatever can be expressed in a homogeneous graphic form—that is, what is suitable to the entire area covered by the map; and to find such a form and to draft it up. Such is, in a brief summary, what the author of the map attempted to do in 1912, as well as in 1922.

The legend consists of about thirty captions and colors. A first subdivision was required: tabular areas versus folded areas. To the former corresponds a spectrum of light hues presented like those of an ordinary geologic map; to the latter, darker hues with four major subdivisions that correspond, as mentioned earlier, first to the group of Precambrian orogenic cycles, then to the Caledonian, Hercynian, and Alpine cycles. The vividness of these colors reveals immediately the appreciably folded areas. A tectonic map should first tend to classify the folded elements according to the age of their movements; then the structural lineaments should be shown, namely, the static tectonics. This is the mode of expression that I have chosen after a very detailed study of a variety of conceivable graphic arrangements.

In regard to the first point, the ideal would be to indicate by means of superposable colors all the phases of all the orogenic cycles that have affected each point of the investigated area.[6] This is out of the question at present; even in the case of a completed scientific investigation, such a presentation would create great difficulties in printing as well as in reading the map, not to mention the numerous simplifications required by the scale. In my final solution (182) of the problem, I have naturally used an approximation and also have abandoned any graphic expression of this kind of complexity. Finally, I have resorted to a classification of the objects according to the age of their major folding or of their most apparent folding, neglecting phases and retaining only orogenic cycles. Even within the limits of these conventions, the graphic expression implies some reservations due to the imperfect knowledge of many areas and to the related difficulty of reaching a final decision. The legend of the map expresses these reservations in its own way; an explanatory text would do so in a more complete fashion, and I shall point out a few of them. There is no doubt about the amount of imperfection that remains in this presentation, but only two alternatives were available: either to drop everything or to go ahead, and whatever feelings I might have, it seemed to me more honorable to face the difficulties and the risks inherent in such an attempt.

Such a long effort carried on for many years, involving the critical study of thousands of geological maps and papers and of tens of thousands of cross-sections, would not have borne fruit had I not received, at different junctures, the most valuable encouragement.

Emmanuel de Margerie, to whom I showed the original map of 1912 shortly after its completion, was kind enough to submit it to a thorough critical examination and to mention it later on to the congress in his talk in Toronto on the geological map of the world.[7] Since then, E. de Margerie searched for ways and means to insure the publication of the map. When, in 1920, it became obvious that the progress of geological exploration required an updating of the document, it was his persuasive influence, as well as the ideas and lists of references that only such a prince of bibliography could provide, that caused me not to postpone, perhaps for ever, such a great effort.

Maurice Lugeon, to whom tectonics owes so many bold and innovative interpretations, has shown on many occasions a great interest in the work in front of you. And the benevolent sympathy of this enthusiastic master, who used to tackle with ease the most fundamental problems, has sustained my efforts more than once.

The organizing committee of the present session has in its circulars drawn attention to the map and expressed the interest that its publication would present; I owe the committee the expression of my sincere gratitude.

This is not all; at the last minute, blind fate almost destroyed the work that had been maturing for such a long time. A few days ago, at the time I was leaving my country for the congress (183), a technical accident happened that, without affecting the scientific value of the map, nevertheless ruined its appearance to the extent of making it essentially unsuitable for presentation here. When informed, Armand Renier, general secretary of the congress, acted with remarkable speed, and I am deeply obliged to him. The restoration of the map was done in Brussels between August 7 and 10, with infinite care, by the Military Cartographic Institute. I am very anxious to express my gratitude to Colonel Seligmann, general director of the institute, for having given all the necessary orders; to the officers of the organization who supervised this most delicate project; and to the civilian personnel who undertook it with skillful hands accustomed to all the fine details of their art.

III. Short Description of Precambrian and Caledonian Asia

What were the destinies of Asia through time? Let us enter upon this history, which for a long time to come shall include blank pages and obscure features. Let us by means of the concepts just discussed try to fill some of the blank pages and shed some light on a few of the obscure features.[8]

Here is peninsular India, which really does not belong to Eurasia from a tectonic viewpoint. With the exception of this gigantic promontory, Asia does not include any large Precambrian massif that has been outlined all along its margin or analyzed exhaustively with respect to the age of the movements. For several parts of China, Bailey Willis has revealed the precise succession of the Precambrian orogenic cycles: this key will be useful, *mutatis mutandis,* for many other Precambrian areas of Asia. While waiting for the time when we shall know about these important phenomena over areas comparable in size to those that have been unraveled in the Baltic shield and in the southern portion of Laurentia, I shall limit myself to sketching the major features that stand out from all the available works.

The Siberian massif, although generally overlain by Paleozoic tablelands, nevertheless displays its Precambrian core in its northernmost portion, the basins of Anabar and of Khatanga. This massif limits to the northwest the Taymyr arc, recently outlined by H. G. Backlund. This arc, which probably extends into the north-south trending islands discovered a few years ago north of the Chelyuskin Cape, is convex to the southeast and has been most probably pushed in that same direction, that is, toward the Siberian massif; besides, the age of this folding remains unknown. In the northeast and in the north, the Siberian massif is bounded by the Alpine arc of Verkhoyansk and by its extensions or related blocks located west of the mouths (184) of the Lena and along the lower course of the Olenek. There are many reasons to believe that this Alpine arc is related outside Asia to the recent foldings of Svalbard and to those that have been recognized in the extreme north of the American Arctic areas: Eureka Sound and the lands of Ellesmere, of Grinnell, and of Grant. Thus one sees the lineaments of a *Periarctic Alpine chain,* often indicated by folded marine Triassic: this Arctic geosyncline is therefore bordered on the American side, as well as on the Eurasian side, by the chains generated from it—chains that in their present arrangement appear back to back somewhat like the two wings of the double Mediterranean chain. This Alpine complex, to which perhaps belong, between the arc of Verkhoyansk, the Arctic Ocean, and the Bering Sea, certain Laramide and Andean elements extending from America to Asia, most certainly forms a large portion of the extensive northeast peninsula of Asia, which is so poorly known today that nothing more can be said about it. To the southeast, the Siberian massif extends beyond the Lena, in the Aldan regions, as a large spur overlain by Cambrian and in some places by horizontal marine Mesozoic deposits. In the extreme southeast, the Paleozoic begins to display folds as it approaches the crystalline[9] of the Stanovoy, and folded marine Upper Triassic occurs west of Okhotsk, in the vicinity of the 140th meridian.

The huge Precambrian promontory called the Sinian massif, after having been entirely mapped in Korea; explored or mapped over large surfaces in the Liaotung and in southern Manchuria; and explored by traverses or sampled during the detailed mapping of small areas of northern China and elsewhere, begins gradually to reveal its complexity. Its limits, in the directions of the northwest, north, and northeast, toward the Greater Khingan or the Gobi, or toward Russian Asia, are still today very far from being established with precision. The Paleozoic, the Mesozoic, and often the Cenozoic covers that subsist in many regions—as patches of moderate extent, or as narrow bands as in Korea, in Manchuria, and in northern China, or as large surfaces as in Shansi, Shensi, and Ordos—appear more and more as objects that were folded at different epochs, and the part played by the Alpine cycle in these deformations is certainly very appreciable.

While the horizontality of the Cambrian and Silurian tabular areas over extensive surfaces renders the existence of a Siberian massif folded in Precambrian times sufficiently acceptable, and while such proof is provided in many parts of the Sinian promontory, this is certainly not always so, for the time being, for the other Precambrian massifs that are supposed to have existed in Asia north of the Tethys. After so many precise observations made in young chains, nobody believes any longer that the metamorphic facies should be a sufficient reason to attribute any rock to the Precambrian: the Caledonian, Hercynian, and Alpine cycles have generated, in Asia as elsewhere, thick series of metamorphic schists. A discrimination is therefore required, and it can be accomplished only here and there. This point being made, let us look northward. (185)

The folds of the Angara beds in the middle of the Siberian shield belong obviously to the Alpine cycle: a portion of the Russian Altai is of Hercynian age. Besides these features, there remain the marginal folds of the amphitheater and the broad area of ancient appearance called primitive crest. I think it wise to abandon the adjective "primitive," which somehow implies an exclusively Precambrian folding, and I shall use the word "crest," or "crests," excluding any hydrographic implication, to designate a large region with incompletely understood foldings that includes, among others, the Alatau of Kuznetsk, the western Sayan, the eastern Sayan, Transbaykalia, and a portion of the Amurian lands, the Khangay and the chainlets of northern Gobi, the Tannu-Ola, the horsts of the Valley of the Lakes, and the Mongolian Altai—all objects that, except for the Alpine movements that reinvolved them later, display within their old frames[10] foldings of different ages, not defined everywhere, and whose extent could not be established without numerous new observations. In spite of excellent preliminary investigations, the documents presently available do not allow us to carry very far any attempt at defining such foldings. However, the known facts are sufficient to indicate the heterogeneity of the old frame, which includes in some places, particularly in the southeast, Hercynian foldings; in some other places, perhaps Caledonian foldings; still elsewhere, Precambrian foldings that seem to have been active in more than one place and during more than one cycle. Therefore, the word "crest" applies to a complex of tectonic units; it is temporarily useful and should be replaced by a more precise term as soon as the state of our knowledge will allow it.

Thick series of semi-metamorphic or clearly recrystallized sediments, associated with granites and other igneous rocks, build the major portion of the crests; remnants of Paleozoic, whose extent and relationships with the major constituents are far from being always well known, occur in certain places.

At this time, the pre-Devonian age of a large portion of the crystalline material of the crests is a positive fact: it is known that the Middle Devonian transgression, accompanied in places by remnants of Upper Devonian and Dinantian, covered the Tannu-Ola and came very close to the Ubsa-Nor, and perhaps reached the upper Kobdo, or maybe even the Altain-Nuru.

The remnants of the Tannu-Ola certainly were related to the formations of the same age as those of the Minusinsk region, rather gently folded and sufficiently framed to suggest a cover folding probably of Hercynian age that would have taken place over an old pre-Devonian basement, a subsurface extension of the crest. In the present state of geological exploration, one cannot say whether the Devonian remnants of the Tannu-Ola are related eastward to the Devonian of the vicinity of Urga; and through it to the zone of clastic sediments, at least partially Devonian, that stretches along the two banks (186) of the Onon toward Nerchinsk, Strietensk, and the Argun; and also to the similar deposits that cover large surfaces in the Amurian lands, beyond which one still finds the Upper Devonian of Ayan on the shores of the Okhotsk Sea. The assumed western extension of the Devonian of Urga has been found along the northern rim of the Dazpkhyn trough, between the 101st and the 99th meridian; beyond, in the direction of Uliastay, around the 98th meridian, are dislocated limestones considered Carboniferous. These Devonian remnants of Mongolia, sometimes horizontal but most commonly dislocated and even in some places involved with high dips in the reactivations of the crystalline basement, demonstrate movements whose age it is not always easy to establish with precision. If one were allowed to use an analogy with the Amurian lands, where the Angara beds overlie sometimes unconformably similar deposits, one would be led, for both regions, to the hypothesis of Hercynian foldings. Besides, it is very probable that the powerful Alpine reworkings, which have taken place throughout all old Asia, are responsible, here as elsewhere, for a portion of the less accentuated foldings that are displayed by the remnants arranged as coverings.

The pre-Devonian age of the crystalline material, sufficiently demonstrated for the surroundings of the Minusinsk region and for a portion of the Tannu-Ola, could also be accepted for reasons of continuity or analogy and, until further notice, for appreciable portions of the crests.

The marginal foldings of the Siberian amphitheater are of undetermined age. Suess did not commit himself on this particular point. Besides, he considered all the crystalline rocks of the crest as Precambrian. In the opinion of L. de Launay, the marginal foldings would be Caledonian, and the same interpretation would apply, over a variable width, to the first crystalline belts of the crests adjacent to the marginal folds; the remaining portion of the crests, along the convex periphery, would be Hercynian. J. H. W. Ahlburg has dogmatically defended the Precambrian age of the entire folding of the crests; according to him, the marginal foldings of the Cambrian and of the Silurian would be of posthumous character only.

The study of the original Russian memoirs, even undertaken with the greatest attention paid to the elements that should be taken into account in this kind of all-encompassing view, mainly demonstrates that one should learn to wait. The energy and the talent dedicated to these Siberian explorations have brought to science numerous, admirable, and definite enrichments. Nevertheless, many facts of fundamental importance remain to be discovered. The difficulties of a general interpretation originate today less from the wide spacing of the traverses than from the small number of significant unconformities discovered. I do not speak of the inferred and graphically constructed unconformities, which should always be considered with caution. The degree of precision of the stratigraphy within the vast complex of the crests, and even within the normal Paleozoic of the margins and of the Siberian plateau, often leads to a wait-and-see attitude, (187) and a lot of work remains to be done to unravel the reciprocal relationships between folded belts. The cautious attitude adopted by so many eminent exploration geologists in regard to problems raised by the chronology of the old foldings presents, for the time being, an obstacle to interpretative attempts.

I shall of course avoid a thorough discussion of all the known facts because its length would extend beyond the limitations imposed by this talk. Besides, a portion of the Russian works printed after August 1, 1914, has remained out of my reach. I shall therefore limit myself to a small number of glances.

The Precambrian age of the folding of certain portions of the crests has been established under conditions that lead either to a temporary suspension of judgment or to a convincing view. To the west, W. A. Obrutchev has been describing for a long time the transgressive character of the Georgian limestone of Torgochino on the old basement of the crest. A few years ago, Cambrian limestones were recognized on the eastern margin of the Minusinsk basin, in the direction of the Sayans. And limestones, of which the Cambrian age is less certain, have been observed along the eastern margin of the same basin near the Alatau of Kuznetsk. To the northwest of Lake Baykal, near its northern end, the crystalline schists can be subdivided, according to M. Tetiaev, into two series separated by an unconformity. In a few places of the region west of Lake Baykal, normal sediments, perhaps Cambro-Silurian, overlie transgressively the old metamorphic rocks. One usually related to the main part of the crests the ancient-looking masses that appear aligned north-northwest, mainly on the right bank of the Yenisey, from upstream of the confluence with the Angara to the vicinity of the middle Tunguskha. In their northern part, these masses are, according to L. Jatchewsky, clearly overlain unconformably by Lower Paleozoic. In their southern part, according to A. Meister, they were folded before the Silurian or before the Cambrian, and an assumed unconformity is supposed to divide them in certain places into two complexes.

A combination of all these data indicates a Precambrian folding that occurred perhaps in two cycles; it does not exclude the hypothesis of Paleozoic reactivations that may have taken place in certain areas. J. H. W. Ahlburg, who defends the strict Precambrian concept, is compelled to move to a rather low position, within the thickness of the complexes described by A. Meister, the lower limit of the Cambrian. But this hypothesis, which restitutes to the Cambrian thick masses of folded and sometimes slightly metamorphosed sediments, appears more prone to reinforce the idea of an important post-Cambrian folding rather than to dismiss it.

Inside the great bend of the Lena, the relationships between the crest and the margin consist generally of an overturning of the former on the latter while the plan is convex northward. In the far east, in the valleys of the Tchentchi and of the Little Patom, the two complexes are conformable according to A. Meister. In certain parts of the basin of the Little Patom, (188) the normal sediments and the metamorphic rocks belong, according to V. Kotulsky, to the same formation. In the valley of the Tonoda, a tributary of the Great Patom, sediments that are almost normal in character grade southward, according to P. I. Preobrajensky, into the metamorphic rocks of the crest; A. Guerassimov is led to consider the latter to be of Silurian age. These observations collected between crest and margin seem, therefore, not favorable to the concept of an exclusively Precambrian folding of the crests; yet, they tend to show that a portion of the margin and a portion of the crest are building, on a large scale, a single unit that is affected by post-Silurian foldings. Certain conglomerates that have been considered as an argument in favor of the strict Precambrian concept indicate only, on the basis of the collected facts, that granite became exposed at a distance, somewhere inside the bend outlined by this portion of the crest, and at an undetermined time. As long as these conglomerates shall not be seen resting unconformably upon their original substratum, they cannot be considered as basal conglomerates, and the possibility remains open for many other interpretations.

It is obvious that if the concept of lateral gradations between crest and margin were to be confirmed, even in a few places, important portions of the first would become part of the second, as foldings of the same cycle. Furthermore, if such parts of the crest were to be pre-Devonian, then the cycle involved could be a Caledonian cycle only. The major portion of the folding of the crests, previously prepared through Precambrian cycles, would have continued during Caledonian times and would be marked by a closing episode that would have reached the first margins and a portion of the crests. This portion of the crest would therefore be the internal, metamorphic zone of a dissymmetric Caledonian chain pushed toward the center of the Siberian nucleus; its margin, excluding subsequent reactivations, would be the external zone, with normal sediments of the same chain. Obviously, the known facts do not allow us to go so far, nor do they sustain the Precambrian thesis in its absolute form.

The Caledonian hypothesis applied to the folds of the marginal Cambro-Silurian cannot, in the present state of our knowledge, be set aside as easily as has been suggested. Indeed, the upper age limit of these foldings cannot be correctly established today, but the unconformity between Silurian and Cambrian, described by D. I. Mushketov, on the left bank of the Lena, opposite the mouth of the Great Patom, has to be taken into consideration. This feature is the only one that indicates with precision, for the time being, a movement belonging to the Caledonian cycle. The upper age limit of the granite of the Krouglaya, near Nizhneudinsk, a granite that intersects and metamorphoses folded rocks attributed to the Lower Paleozoic, remains unknown. While we wait for additional observations of a more decisive nature in one direction or another, the Caledonian hypothesis remains plausible, but that is all. (189)

In the southeastern region of the crests, investigations carried out in recent years have led several times to the discovery of crystalline schists of Devonian and Dinantian age. In the basin of the Gazimur River, north of Nerchinskiy Zavod, V. Zveyrev has reported crystalline schists that result from the metamorphism of fossiliferous Neodevonian and Dinantian. In the basins of the Never and the Oldoy, the left side tributaries of the Amur between the 123rd and the 124th meridian, P. Kazansky has observed a complete Devonian section and some Dinantian intruded by granite and partially changed into various types of crystalline schists. The same author had previously reported fossiliferous Ordovician in the basin of the Omutnaya, west of the 123rd meridian. In the basins of the Oldoy and of the Oldokon Y. A. Makerov has found metamorphic schists of

Devonian age. These discoveries confirm the views of L. de Launay, who has so firmly opposed the exclusively Precambrian age of the crests. In the territory explored by P. Kazansky, a granite of the Alpine cycle has intersected Angara beds. The metamorphism that has generated the metamorphic facies at the expense of the Devonian and the Dinantian along the zone occupied by these deposits remains, nevertheless, most probably, essentially Hercynian, and it is obvious that it could not be any older. But as long as the relationships between these crystalline schists and those of earlier cycles are not better known, it would be rash to delineate in the crest any areas or zones corresponding to the foldings of the different cycles.

At present, four points appear sufficiently demonstrated: the folding of a portion of the crests is older than the Cambrian; a large but undetermined part of the folding is older than Middle Devonian, without further precision available; the marginal folding is younger than some Late Silurian beds; the folding of certain portions of the crests, mainly in the southeast, is of Hercynian age.

If I were asked which working hypothesis would temporarily be most appropriate to this large problem of the crests, I would answer that it would be the hypothesis of a predominantly geosynclinal regime, with foldings having occurred during several cycles as follows: the first ones are of Precambrian age and in part at least Algonkian; certain portions of the geosyncline were closed by folds before the Cambrian; other parts were preserved or have reopened during one or several subsequent cycles, the last of which was the Hercynian cycle, during which the old frame reached completion. This hypothesis, with its numerous mechanical artifices and its unlimited flexibility, seems appropriate to a situation consisting of a few accepted facts and of many doubtful ones, a situation characteristic of a problem in which several of the elements essential for a full appreciation are missing. I shall neither attempt here to develop the possibilities offered by this hypothesis nor try, at this moment, to eliminate those that are unacceptable. (190)

In a few minutes, I shall deal in an appropriate manner with the deformations undergone by the crest and by the margin during the Alpine cycle.

Farther south, pre-Devonian fragments much smaller than those of the large Mongolian and Siberian crests are involved in unknown number, and with boundaries established in a few places only, in the foldings, essentially of Hercynian age, of the Kunlun. It is obvious that these old fragments, brought to light by Alpine foldings, together with the Hercynian folds they underlie, demonstrate pre-Hercynian cycles: a future problem is to detect among them what could be Caledonian or Precambrian. The large ancient massif that has been postulated as one single block in the middle of Tibet is not confirmed by recent investigations, and concerning the areas of Precambrian folds that have allegedly been detected several times in southern China, the most up-to-date sources show mostly what remains to be done. In the mountains east of Mandalay, the accurate explorations of T. D. La Touche have disclosed an unconformity between Gothlandian and Ordovician, the clear record of Caledonian foldings involved in more recent folds.

The fact that the Caledonian foldings had in Asia a much greater importance than shown at present by the most direct evidences is indicated by their action, first on the Precambrian continental areas, second in some geosynclines that were supposed to close only during later cycles. The uplifting shown on the Siberian platform by the Ordovician gypsum-bearing beds and the emergence revealed over a large portion of the cover of the Sinian massif by the dolomites of the uppermost Ordovician and by a lack of Gothlandian—which seems of such general nature in northern China, in Manchuria, and in Korea—both indicate movements with a large radius of curvature and with an upward component that can under no circumstances be attributed to epeirogenesis: they are actually very flattened *basement folds* that the Caledonian orogeny generated in the Precambrian frames, and consequently in their cover, in the same way that the Alpine orogeny did later on in all kinds of basements with old folds. In respect to the Sinian massif, and in the same regions, there are in addition some similarly flattened basement folds that were generated by the Hercynian orogeny because Devonian and Dinantian are missing, more or less under the same conditions as the Gothlandian. Repercussions within the Tethys are not absent: in the Spiti area, there is an unconformity between Ordovician and Cambrian; in the Salt Range, the great gap that separates with an unconformity the Cambrian from the Carboniferous boulder clay probably results, as the hiatus of the Sinian lands, from the effect of Caledonian foldings followed by Hercynian foldings; besides, it is impossible that one of the two cycles did not operate.

Assuming that the marginal folds of the amphitheater, which are identified or followed from the bend of the Lena, near the 118th meridian, toward Irkutsk, then along the Yenisey (191) and across the lower part of the three Tunguskha, are really Caledonian, then there would be no difficulty in connecting them westward, underneath the Hercynian Novaya Zemlya, with the Caledonides of Svalbard and Greenland, as well as with those of Scandinavia and of the British Isles; a Cambrian and Silurian geosyncline would have snaked between the Fennoscandian massif and the Siberian shield and would have been filled with folds; it would have been the precursor of the Hercynian geosyncline from which were generated the Urals, the massif of the Kirgiz steppes, and the major part of the Russian Altai; furthermore, it would have been the ancestor of a portion of that depressed zone that during the Alpine cycle extended more or less in a meridian direction from the Kara Sea to the sea of Oman, through the Ob basin, the Turgay, the Turan, and Iran—a depression that intersected that of the Tethys and that is crossed today by the double festoon of the Iranian arcs, clearly curved at its crossing because of the lesser resistance encountered there by the Alpine folds; this depression is visible today mostly at its extremities, on the one hand in the Mozambique strait, on the other in the Turanian plains, in the immense Cenozoic and Recent accumulation basin of the Ob, and, finally, in the Kara Sea, framed by Novaya Zemlya and the Taymyr peninsula.

Let us now consider, on the other side of the Earth, the huge surroundings of the Atlantic Ocean: beyond the Arctic, Scandinavian, and British Caledonides are the Caledonides of the Taconic Range and those of the Piedmont; then those of the Sahara; and those less demonstrated by paleontology, but still probable, of Dahomey and of the oriental margins of South America; and finally those that I think we can accept in Laurie Island of the South Orkney Islands, where folded Silurian, trending northwest to north-northwest, cuts across the Andean-Alpine arc of the Austral Antilles—trending here west to east under conditions that suggest the picture of a Caledonian fragment incorporated within that arc and brought to light by foldings of the Alpine cycle. Thus, here is, between the highest northern latitudes to the 62nd latitude south, the acceptable length of the main branch of the Caledonian chain; the geosyncline in which it was born covered huge surfaces; its direction, as that of the folds that rose from it at the end of the Ordovician, at the end of the Gothlandian, and during a few late replicas in the Devonian, is nearly meridian. One can see now that this geosyncline was a first Atlantic, a precursor of the present one; that the individualization of the double American block, on the one hand, and of the ancient blocks of the Old World, on the other hand, goes back an enormous length of time into the past, to the Early Paleozoic; that the welding of these continents was accomplished again, temporarily, by the very effect of the Caledonian foldings, which are in fact so well completed that some Hercynian and perhaps some Alpine chains were able, from Europe or Africa to America, to extend from east to west, over and across the old Caledonian frame; and that the *present* (192) Atlantic, gradually formed in Recent times, is only a second Atlantic with an origin comparable to that of the Caledonian geosyncline. The repetition of such circumstances, with such a crossing of patterns, after such an enormous length of time, and on such a scale, raises problems that we shall not fail to examine later on.

In comparison with this major branch, the problematic Caledonides of Mongolia and Siberia would be a small matter; however, one sees that they would reproduce, to a certain extent, the orientation and history of that branch. At any rate, the destinies of the weakness zone of western Siberia and of the Atlantic show more than a simple parallelism. The first Atlantic existed, at least in broad and long incipient segments, in the Cambrian; the subsidence of the depression between Fennoscandia and

Siberia is expressed in the extreme north by the Ordovician of the Yugorskiy Shar, the Gothlandian of the island of Vaygatch and of the Timan; other small outcrops of Silurian are known very far south in the mountains between Chu and Ili and in the vicinity of Lake Balkhash.

During the Early Devonian, the meridian depression is already the powerful geosyncline from which the Urals will emerge later on, and similar conditions become apparent in some places at the same time in the Kirgiz massif, in the Russian Altai, and in the Tien Shan; during Middle Devonian, such conditions prevail over these areas and from them, as well as from the Tethys, the transgression of that time extends over the old Mongolian and Siberian crests and over the pre-Devonian frames of the Kunlun. At the time of the Hercynian paroxysms, the depression extending from Turan to Novaya Zemlya is filled by folds that in the former of these regions trend obliquely across the furrow in a way similar to that of the Paleozoic folds from Europe to America.

It can be assumed with the greatest probability that the Hercynian chains of Asia do contain Caledonian enclaves that are similar to those occurring in the Ardennes, in the Rhenish schistose massif, in the Thuringian Forest, and in the Lysa Gora. From the Ardennes to the Vistula, it is obviously the not too distant resistance of the hidden southern margin of Fennoscandia which has brought into existence, by generating the tangential effort, the east-west trending branch of the Caledonides.

IV. Short Description of Hercynian Asia

The Hercynian folding of the Urals also is responsible for the massif of the Kirgiz steppes, a portion of the Russian Altai, the Tarbagatai, the Dzungarian Alatau, the Tien Shan, and most of the Kunlun with its branches: Yarkand arc,[11] Altyn Tagh and Nan Shan, with their rear chains, and Chinling Shan. (193) The different branches of the Kunlun contain fragments of pre-Devonian folded frames brought to the surface. Precursor movements were active in many places during the Devonian. A very important phase of folding occurs in the Kunlun and the Tien Shan at the end of the Devonian and at the beginning of the Dinantian; a second paroxysm, which unquestionably corresponds to the intense and most frequent phase in Hercynian lands, affects these two chains during the Late Carboniferous. Replicas occur at places during the Permian. The Chinling Shan, between the 104th and the 106th meridian, displays a typical Hercynian structure with pre-Devonian enclaves. At one location of the Ngan-Hoey, the Permian rests unconformably on Upper Carboniferous: the Hercynian frame reaches the shores of the Pacific. From there to the countries of western Europe, it stretches over more than 11,000 kilometers; belonging to this trend are the 3,500 kilometers of the American segment.

F. E. Suess has followed the Hercynian folds from Asia into Europe, and there is little to be added to the sketch he has made of the western branches. The Fennoscandian massif forms the backbone of most of the Russian platform; a line drawn from Scania to the vicinity of Astrakhan probably marks its southern margin, and the crystalline buttonholelike inlier of Pavlovsk on the Don (in the vicinity of the 40th meridian and the 50th parallel) most probably belongs to it; its eastern margin must run very close to a line that from the entrance of the White Sea follows the margin of the Timan and the Urals and reaches the front of the Mugodzhars. A large promontory oriented toward the southeast thus seems to end it toward the northern Caspian steppes. The perimeter thus outlined may, furthermore, enclose Caledonian elements. This spur divides the large Hercynian sheaf that originated from the Altai and from central Asia. The northern branches bend northward, forming several loops visible in the massifs of the steppes or concealed beneath Recent deposits, to join the Urals and to build, undoubtedly, an appreciable portion of the bottom of the Ob depression. In the extreme east, the folds of the Russian Altai are stopped by the Alatau of Kuznetsk and by the Mongolian Altai, the two salients that the major part of the older crest displays to the west; the Hercynian folds must adapt, by means of various types of inflections, to this double obstacle and to the reentrant that separates them. The major portion of the southern Hercynian branches originating from central Asia stretches to the eastern end of the Precambrian massif of southern Russia, which leads to another separation into two sheaves of unequal importance: the Donetz sheaf, which flows and narrows between the Podolian massif to the south and the Fennoscandian massif to the north, showing some weakness (194) during the Hercynian deformation and the Alpine deformations; and the great Varisco-Armorican sheaf between Dobruja, Finistère, and Ireland, which is too well known to require particular attention. It generates the important branchlet that reaches the Moroccan Meseta and the Saharan Altaides through the loop of the Asturias and the Spanish Meseta. Thus are unfurled, as increasingly divergent twigs, the branches that originated from the common Hercynian trunk of central Asia (Figure 8).

From Minusinsk to the Sea of Okhotsk, across Mongolia and the Amurian lands, in the Devonian deposits accompanied in the west by some plant-bearing Culm, Hercynian foldings predominate. While some of them generate waves on top of the old frame, others are more roughly pinched in the deformations of the crystalline, even to the point of being metamorphosed and intruded by granite.

The phase of the Hercynian foldings that by the end of the Devonian and the beginning of the Dinantian has strongly affected the Kunlun and the Tien Shan has made itself felt in the old areas as basement folds with large radius of curvature and with usually upward vertical effects: in the western and northwestern parts of the crests, the Middle Devonian marine sandstones are overlain by a gypsiferous and a saliferous Upper Devonian, in turn covered by plant-bearing Dinantian. It is the same type of deformation that has allowed conditions of emergence to be maintained in northern China, Manchuria, and Korea during the Devonian and the Dinantian, while the sea returns to cover these lands only in later Carboniferous times.

South of the Yangtze, marine Devonian and Dinantian reappear: the Paleozoic sequence, generally more complete, is also more folded; large ancient fragments lie underneath this sequence in more than one place where the Paleozoic cover is closer to a tabular condition. An enormous amount of work remains to be done to unravel these areas and to understand the style of the Paleozoic foldings that they have undergone and that are of variable intensity.

The Himalayan geosyncline in turn displays, in particular at the base of the *Productus* shales, the vertical effects due to Hercynian influences. Furthermore, I have previously recalled the great hiatus observed in the Salt Range.

V. General Features of the Alpine Cycle

From the Permian or the Triassic until the Present, the Alpine orogenic cycle has developed with an enormous variety of phenomena. In this cycle, the mobility of the objects usually is displayed much better than in the previous ones; however, the relative simplicity of the Hercynian and Caledonian events is an illusion that originates from the fact that we see them in a more condensed and shortened manner. The time has long past when all the history of the Alpine foldings seemed to have taken place during a short interval of the Tertiary. A dynamic tectonics that reconstructs the first outlines of chains; that distinguishes among the different orogenic paroxysms, in the Jurassic, in the Cretaceous, in the Tertiary (195)—phases that stand out among others of lesser importance; a dynamic tectonics that analyzes only in order intimately to relate each episode to the flow of events that generates it and from which it is distinct but not dissociated; that recognizes in the discontinuous either an artifact of dissection or a reality that should not be separated too much from the continuous evolution that surrounds it—such a dynamic tectonics, I may add, gradually eliminates features that less flexible views have considered too well established. A given paroxysm of great importance here, there ranks only as a precursor folding or as a late replica, while the deformations do not necessarily cease to be independent as a whole; in a given chain, arc, nappe, or fold, the components display phase differences, while in each phase it is not at the same moment that the stress reaches its maximum intensity according to the point considered in each object. This is because matter resists deformation and because any resistance requires time. What is true for phases applies also to orogenic cycles, which, after all, are master phases.

With the exception of a few well studied but limited areas, only scattered data are available on the phases of the Alpine cycle during the Triassic and the Early and Middle Jurassic. In a major part of the Andes, an important paroxysm takes place, with phase differences distributed by regions, toward the end of Jurassic times or slightly later; strangely enough, it occurs just before the Portlandian in the region in which this great event is best known, that is, the eastern half of the cordilleras of the western United States; important replicas occur during the Cretaceous in different segments. Therefore, one sees within the Alpine cycle a long phase so important through its major effects that it may be designated without exaggeration as the *Andean subcycle,* with its precursor foldings, which extend at least from the beginning of Alpine times, its paroxysms, and its late foldings during the Early Cretaceous. The sequences that in the western United States were to be twisted into Andean folds had been deposited during Paleozoic times, and within them small gaps, usually devoid of major unconformities, represent the weak repercussion of Caledonian and Hercynian foldings whose major location was obviously the Taconic Range, the Piedmont, and later the Appalachians. The Andean folding, properly speaking, has usually affected formations in these areas that previously had not been appreciably folded and that, once involved in the major folds, had furthermore to accommodate the positioning of huge batholiths. With respect to Laurentia, these foldings remained in an external and marginal position. The Laramide folding, on the contrary, overlaps the western margin of this old shield. The swells of the Rocky Mountains, these broad anticlines in the core of which appears, strongly warped and sometimes complicated by clean-cut thrusts, the Precambrian base of Laurentia itself (196), are the result of an effort during the Cretaceous-Tertiary boundary that added to the folded basement a new swarm of more oriental chains, but this time at the expense of a very old promontory and its unconformable cover. The low plasticity of this old folded basement prevents the generation of a style as compressed as that of the Andean folds: hence the predominance of this broad swell. It is really the heavy and powerful architecture of an almost indurated land, which only a huge folding effort has been able to revive very late: it is a basement folding. The Alpine foldings, properly speaking of Tertiary age, have not created anything entirely new except the Coast Ranges, along the Pacific margin. By reworking through large radii of curvature the Andean frame, also indurated for a long time, the Alpine foldings have led to a new generation of basement folds, such as the ancient basements of the Coast Ranges, the furrows of the Gulf of California, of Sacramento, of Puget Trough, of Cook Inlet, the powerful anticlinal bulges of Baja California, of the Sierra Nevada and British Columbia, of the Saint Elias Mountains, of the Kenai peninsula and Kodiak Island, of the Alaskan peninsula and of a large portion of the Alaskides, the large warpings of the Cascade Ranges, the ondulations of the lavas of the Columbia Plateau, the synclinal subsidence of the Great Basin with its secondary waves, and the infinite complication of fractures, often longitudinal, that are subordinated to this new swell, which is too much indurated to fold without fracturing. These are the features—extended in the same style through replicas during the Late Tertiary and the Pleistocene—that have created the modern orographic conditions of these old things, so different from the ancient, now dislocated plan. Let me point out, furthermore, at a much later time, renewed deformations with great radii in the Rockies, with related morphological effects. After this description, it is time to return to Eurasia.

Concerning Europe, I have recently mentioned [12] how long-known facts become related and shed light on each other as soon as they are interpreted as foldings, infinitely varied in their shape and in their intensity, that operate in direct relationship with the distant Andean paroxysm: in the geosyncline, the deformation of narrow furrows that deepen until the Tithonic, where abyssal deposits occur in them, while nearby are uplifted island festoons bordered by coral reefs and, farther offshore, by a rim of coral muds; in the geosynclines again and elsewhere, the so-called Cimmerian folds of Crimea and of southeastern Europe; still elsewhere, narrow cover foldings like those that begin to appear along the margins of the Harz, or those, much broader, that occur in the Boulonnais, or folds that keep very large radii of curvature such as those of the Purbeck area, which extends from southern England to the Jura across a large portion of France. To account for all these features, it is necessary only to assume orogenic movements. Truly, these foldings occur in Europe during the Portlandian and the Valanginian and still at other times; (197) one could therefore consider them unrelated to the Andean paroxysm since they occurred before the Portlandian. But let us consider the imperfect rigidity of all the media that are involved in tectonics. None of these media allows, over a long distance, the rigid and instantaneous transmission of a thrust: in the continental bottom and subbottom, there are simply semi-plastic or plastic deformations distributed within volumes that influence each other by contiguity and by continuity and that work all together in a synergic play. The surface adapts itself as well as it can to this deep-seated deformation; we call the result folds. The episodes that the heterogeneity of the upper parts generates within this flux have no reason to be synchronous except on a large scale, and in general they should differ somewhat in their age. In particular cases and over short distances, it is possible to visualize synchronous deformations—still this can be true only within certain limits: the low sensitivity of the stratigraphic column with respect to the most delicate tectonic phenomena overwhelms the small differences in the illusion of a perfect synchronism. Furthermore, we see that these small-scale European effects follow the great American folding: the match is excellent. The synergy of the deformations is so much better explained since Europe and North America were forming a single mass in those times.

Of the two faces that Eurasia turns toward America, one, namely, Europe, has been noticeably affected by the Andean deformations; therefore, it is not impossible that the other face, namely, the Far East, has been affected up to a certain point. In fact, the Andean frame that exists in the Alaskan peninsula, and most probably in the center of the Aleutian arc, is not particularly lacking in solidarity with East Asia. In Japan, plant-bearing beds have been reported at different levels within the Jurassic, and other plant-bearing beds form the Lower Cretaceous sequence of Ryōseki. One can interpret the marine intercalations of this Jurassic as an indication of alternating movements, with changes of sign. The high situation, most consistently expressed during the Early Cretaceous, can be considered as the effect of a rather strong Andean phase or of its major late replicas; however, it is not possible as yet to detect anything else but the vertical effect of all these movements.

From the Malaysian straits to Korea, a large portion of the continental lands of the Far East is affected by foldings and by the positioning of batholiths about which we know only that both are later than certain levels of the Lower Mesozoic, and particularly of the Jurassic. The *terminus ad quem* of these events having as of today not been defined in any way, one must be contented for the time being with relating them to the Alpine cycle in general; the idea of an Andean phase, of Jurassic or Cretaceous age, presents itself naturally to the mind (198) although Cenozoic deformations cannot be excluded. It is not wise to go beyond the question of chronology; rather, let us look at the evidence.

In the Malacca peninsula, huge granitic batholiths have metamorphosed Lower Mesozoic continental series that have been compared to the upper levels of the Gondwana series. Smaller granitic bodies, in which similar relationships are assumed to exist, have been reported in Thailand. In French Indochina, the beautiful synthesis of C. Jacob has very recently revealed huge overthrust nappes that have involved all the formations up to the Liassic;[13] thrusted to the southwest they have covered the entire Tonkin and overlapped the northern part of Annam. Mesozoic rhyolites have participated in the movements, and porphyrites frequently occur along the major contacts. Since the amplitude of these nappes approaches 300 kilometers, one cannot doubt their extension over large areas of southern and southwestern China, where they must play a very important role.

Almost everywhere in China, in the south as well as in the north, large or small patches of Permian, Triassic, or Jurassic have been involved in foldings that should be related to the Alpine cycle. The movements of the southern margin of the Chinling Shan, considered as Permo-Mesozoic, may belong here: they have affected the lower part of the Angara beds, which are folded, intruded by granite, and overlain unconformably by other Angara series considered Jurassic, which furthermore have been refolded later. Therefore, here again it is not possible to establish an upper age limit for the entire folding.

In Korea, we can be guided by the numerous works of our Japanese colleagues that have found their almost final graphic expression in a superb document, the Geologic Map of Chosen, at a scale of 1 : 1,000,000. There is no doubt that the four great cycles have been active in Korea, not to mention the strongly granitized Precambrian foldings. The Caledonian and Hercynian movements have been active, as mentioned previously, as in the remaining part of the Sinian massif through very large basement folds that have maintained conditions of emergence from the Gothlandian to the Dinantian. Such phenomena do not occur without beds displaying in some places much higher dips. Furthermore, tight folds are frequent. The Paleozoic covers that form ondulating large patches in southern Pyŏng-an-do and in Hoang-Hay-do—and elsewhere as smaller remains—owe a part of their aspect to Paleozoic foldings, namely, Hercynian. In southern Hoang-Hay-do, Angara beds overlap transgressively truncated folds that include Ordovician as well as Carboniferous or Permian formations. The Angara (199) beds of the vicinity of Pyŏngyang, themselves dislocated by Alpine deformations, lie unconformably over Precambrian and Lower Paleozoic; this is also true for the Angara series of the basin that extends southeast of Chinnampo.

Southeast Korea, in the two Kyŏnsan, displays an extraordinary development of strongly folded Angara beds of Jurassic age intersected by granitic batholiths and accompanied, as in Tonkin, by great masses of porphyries and, particularly, porphyrites. Important Alpine movements, and perhaps Andean, have occurred in this area, and the reappearance of an association, very comparable to that which distinguishes the nappes of northeastern Indochina, is a fact to be stressed. In the south and southeast of Korea, the structural trend is north-northeast: a broad zone of gneiss limits the Jurassic of the Kyŏnsan to the West; beyond are located the narrow Jurassic alignments of the northern Chŏlla-do and, along the same trend, the Paleozoic patches of southern Kangwŏn-do, to a great extent pierced through by granite; more to the west, there is another zone of gneiss that in the two Ch'ungch'ŏng is limited further on by new Jurassic alignments, so narrow that one is led to think that they are strongly pinched scars or synclines deeply engulfed in the crystalline. Therefore, all that extends from here toward the southeast, until the Korea Strait, has appreciably been affected by Alpine basement folding. What extends beyond, namely, central and northern Korea, is a country with a similar basement constitution affected, as all Eurasia, by Alpine deformations but in a less intense manner than the southeastern Korean lands: it is a relative foreland. The fact that the most important of these Alpine movements originating from the southeast do stop along the line of the Ch'ungch'ŏng, or of the Chŏlla, or in the Kyŏnsan is a matter of detail that can be omitted.

Granites have pierced the Paleozoic covers of the Pyŏng-an-do and of the Hoang-Hay-do. In certain places, or over large regions, Alpine granites may have affected Precambrian rocks only; therefore, one can understand the reservations that still exist—and will continue to exist for a certain time at least—concerning the age of parts of the granites in adjacent and less-known areas such as Manchuria and Sikhota-Alin. Besides, there is no doubt that the Precambrian basement predominates in these areas over large surfaces, as in Korea. One can also perceive how the Alpine dislocations that have affected to a variable degree all China and particularly southern China, framed in at both ends by Tonkin and Korea, more often than not resemble those of the two latter countries; and one can also visualize the increasing importance taken on by these foldings and thrustings, completed, so to speak, in a dry state, without geosyncline but nevertheless granitized.

The patches of marine Triassic in the vicinity of Vladivostok, of the Ussuri, and of the shores of the Okhotsk Sea facing the islands of Shantar demonstrate Alpine movements whose upward age cannot be precisely established as yet. (200)

The Andean deformations have not been limited to the Atlantic and Pacific faces of the Old World: all of Eurasia has been delicately modeled by them. The middle part of the continent has not failed to react in different places. In the Salt Range, Lower Cretaceous overlaps transgressively Middle Jurassic: therefore, a partial geosyncline of the Tethys has moved upward. The Spiti shales of the Himalayas, which, between the neritic deposits of the Middle Jurassic and the Lower Cretaceous Giumal sandstones, indicate a phase of deepening of one of the major furrows of the Tethys during Late Jurassic and Early Cretaceous times, show an opposite movement, but all this is but folding: cordilleras rise and furrows sink. Stratigraphy reveals only the vertical aspect of these movements; tectonics adds to it the horizontal effort without which nothing of that sort would happen. The dynamic analogy with the arcuated embryos of the Alps, at the same time, is striking. The broad depression from the mouths of the Ob to the Turan underwent deformations with large radius of curvature. Along the Lena, between Yakutsk and the Arctic Ocean, the Upper

Volgian and the Valanginian filled a broad curved furrow, which is but the foredeep of the Verkhoyansk arc: the deformation of this cordillera and of this foredeep is, therefore, obvious since that time. This situation required some uplifting of the center of the Siberian massif around which the sea had to bend toward the north in order still to be able to penetrate, toward the south, into the Ob depression, where it left traces of Volgian age in the vicinity of Obdorsk and of Valanginian age, almost as far as the 62nd parallel.

Let us now examine the Laramide foldings, this great American preface to the pre-Lutetian movements so widespread around the world, particularly in Eurasia. The folding of the Rocky Mountains, accomplished during the Danian and probably also during Early Nummulitic times, preceded somewhat the bulk of the pre-Lutetian movements. There is, therefore, between these two events the same kind of synergy as that between the Andean paroxysms and the Andean deformations of Eurasia; this is further shown by the fact that the tonnage of the true Andes, as well as that of the true Laramides, is noticeably greater than that of the other Andean or pre-Lutetian deformations. The well-defined emergence that prevailed during the Early Nummulitic over such a large portion of the Alpine chains, particularly in the Alps, therefore appears under a new light: there is synergy with the greatest deformation that opens the period; there is delay in deformation with respect to the same event. There is, as during Andean times but with more decision and more energy, at least in the Alps, an increase in the tangential effort in the furrows and in the cordilleras already beyond the embryonic stage; there is culmination of the whole and the Maestrichtian sea is pushed far away. When this little paroxysm, the precursor of many other deformations, decreases its horizontal effort, everything subsides and the sea advances: it is Lutetian times. And, if this effort has shown in the Alps (201) more power than the Andean effort, the same dissimilarity has occurred on a much greater scale in the United States: the Laramide tonnage is of the same order of magnitude as the Andean tonnage, but it consists of already indurated material, therefore much more difficult to deform.

VI. Behavior of the Tethys under Compression

At the time of compression, that is, during the paroxysms and during more than one phase of precursor times or even of replicas, there is, either everywhere or in certain segments, a very great predominant deformation as well as two derivative effects of unequal importance.

The predominant deformation—the Pacific backsides of Asia, Australia, and the two Americas, in a word, the domain of the Circumpacific chains being left out for the time being—this predominant deformation is the *drawing together* of two continental jaws, Indo-Africa and old Eurasia with a narrowing of the Tethys. With respect to the double front displayed by the two opposite jaws and with respect to the oriented stresses responsible for the narrowing and acting either on one of the continental blocks, or on the other, or on both simultaneously, these blocks acted as *rear zones:* I shall for a few minutes use this kind of language. The tangential stress varied within each jaw according to sectors; the resistance on both sides also varied according to the degree of rigidity and as a function of the sinuosities and salients of the margin; *the rear zone* of each sector, in either jaw, participated, to a degree that we shall have the opportunity to define more precisely, in the same kind of deformation as the frontal sector itself. Therefore, there is rather early a first distribution of the system—jaws and geosyncline—in segments of very unequal lengths and of a width that increased with time toward both rear zones in proportion to the progress of the compression. Each of these segments was devoted, within the overall deformation, to a particular task, which became increasingly specialized over time. Finally, the action of the horizontal stress on each of the transverse alignments varied with time: it increased, decreased, and increased again, and so forth in alternating episodes of very unequal duration. In that respect, I should stress two points that have fundamental consequences. First, tectonics has brilliantly and at length dealt with segmental variations of resistance and with their effects on the history of arcs. However, the problem of the spatial variation of stress, that is, the tangential effort that varies, at the same time, from one transverse alignment to another, has not received the same attention; nevertheless, it is of fundamental significance and certainly no less important than the problem of resistances or of variations of stress in time. Second, segments and transverse alignments cannot be, as soon as great volumes and long distances are involved, rigidly defined segments and transverse stress alignments; deformation always is somewhat plastic, (202) slightly more so on the average in the geosynclines, slightly less so in the continents; the curved plan that characterizes all Asia, and for that matter the entire Earth, should be sufficient to make us aware of this situation while waiting for deeper insights. Therefore, there are, properly speaking, only *transverse flux alignments* and *flux segments* characterized by a certain regime of flow and deformation.

The greatest of the derivative effects is the deformation of the jaws including the rear zones, or, in other words, the deformation of the continents. The movement of a jaw—and in a more general fashion of any continental block—cannot occur absolutely as one single mass, even when this displacement takes place horizontally against a less resistant medium. This situation results first from the fact that plasticity within the mass of the continent varies as a function of the bathymetric level, from the subsurface down to the bottom and the subbottom; second, it results also from the heterogeneity of the upper parts, which consist of compartments so dissimilar that each one reacts differently to the general plastic deformation; finally, there are additional factors. Thus, what will be the distribution of energies in a broad schematical way? In general, a large portion of the original energy will be consumed by deformations within the continental block itself; it cannot be otherwise because this derivation of energy is the *first one* that can be discussed at present with some degree of reliability; this consumption occurs *in situ*—it involves volumes in comparison with which all the geosynclinal chains, and other types that emerge from the seas, weigh very little, at least in the Alpine cycle. The collision of two continents, or the more gradual confrontation between the two portions of the same continent separated by a zone of weakness, is expressed by a certain repercussion of the compression on the jaws themselves. Therefore, within the width of the colliding objects, a kind of backlash occurs whose deforming effects—all other features being equal—gradually decrease from the zone of collision toward the rear areas: in the case of very powerful compression, this *second* derivation of energy may become extremely efficient. Thus, the first derivation, complicated or not by intense secondary compression, always produces within the mass of continents very powerful effects. As we shall see soon, these effects consist essentially of a folding, with a large radius of curvature, of old indurated frames: these *basement folds,* through the internal tensions resulting from their formation, generate important fractures. Within continental areas, fractures will most probably not occur unless such conditions exist or unless there is distensional traction.

The other still important derivative effect is, in the case of a vise, the folding of geosynclinal chains: arcs are formed whose distribution and deformation depend, originally, on the first fragmentation, to which is added later a division in much shorter and more specialized segments, under the influence of the rupturing through fractures that reaches the jaws and divides them into high and low sectors of unequal resistance, not to mention other factors that will become more evident later. (203) The amount of energy diverted by this *third* derivation to the benefit of geosynclinal chains and of other new chains can increase, particularly when the jaws have a small amount of heterogeneity and are therefore not much deformed.

Geosynclinal chains usually are overturned *backward,* that is, toward the jaw that is more directly responsible for their formation. When they happen to be in close contact with the jaw or to be thrust over it, these chains restitute to the upper parts of the continent a small portion of the energy they acquired from it.

This restituted energy, a *fourth* component of the energies in a distributed state, is naturally very much lower than the others. Its effects usually are limited to that part of the continental margin that is most exposed to the backward pressure of geosynclinal chains. The restituted energy propagates from this margin toward the rear zones, and its effects decrease rapidly in that direction. The propagation within the continental block takes place through the unconformable covers or the basement consisting of old indurated folds. In the first case, *cover foldings* are generated; in the second, the energy restituted to the old frames is added to those energies that are liberated directly within them and that participate together in the generation of basement folds and related fractures.

In the distribution of the useful energies that occur within a continent and in its related blocks, it is therefore possible to visualize at least four derivations more and more remote from the original energy. The first one, more important than all the others except in particular cases, is never absent as long as there is any trace of movement of the continent. The second is a consequence of compression; it is really only a portion of the first one and is active similarly inside the continent with a few peculiar features concerning its location. The first, including this secondary derivation whenever it exists, is the energy that can properly be called *intracontinental*. The third, which is borrowed from the first or, which amounts to the same, from the second, remains very low compared with the intracontinental energy; it is consumed in new chains of geosynclinal or Circumpacific type. The fourth, still much lower, is the energy *restituted* to the continent, and in a detailed viewpoint there are other kinds of distribution. When discussing East Asia, we shall indicate to what extent these factors have a chance to occur in the deformation of the Circumpacific chains and of the five continents against which these chains are molded. Meanwhile, we shall look at the Tethys and its two rear zones.

In tectonics, the prevailing practice is to relate the direction of movements to continents assumed to be motionless, to old massifs considered in their passivity as obstacles and not in their active displacements. The rear zones correspond to the forelands of this older nomenclature, (204) and in the remaining part of this work, I shall try as much as possible to use this manner of speaking. The other terminology, as may be seen, has the advantage of better defining the origins, the deformations being expressed in terms of power and not of resistance. It is clear that the fact of the nearing of jaws—with consequences drawn or to be drawn—remains independent of any convention. With respect to initial energies, the media and the conditions under which they are liberated escape our visualization so much so that we are not going to say anything more about them.

VII. Behavior of the Chains of the Tethys

At the time of paroxysms, the chains of the Tethys were deformed with the greatest variety of movements and shapes. Today, it is not possible to describe fully, as much as the testimonies would allow, the story of this fundamental event. It would be necessary for each chain or element of chain to determine the mechanical conditions imposed on the arcs by the original segmentation, by the shape of the settings cut inside the ante-Alpine frame; to indicate for the cordilleras and the furrows born in them, the initial features of embryonic tectonics; I mean the kind of folds that once generated can no longer disappear but only become more accentuated or undergo fragmentation. It would be necessary to unfold—going back in time—stratigraphies presently rolled up into folds and to reconstruct sea bottoms; to appreciate, on the basis of facies changes, the importance of the small vertical movements and put them back within the tangential effort that regulated everything; in summary, it would be necessary to reconstruct the deformations due to precursor foldings and to follow them until the eve of the paroxysms whose times change according to locations. These preliminary conditions being fulfilled, one could consider the main problem; one would see how all these factors, whose effect could not be entirely appreciated until completion of the Alpine cycle, can impose themselves, through a heritage that cannot be repudiated, upon the paroxysmal deformation itself. One would see, for instance, how during paroxysms phase differences can persist that the precursor foldings had more weakly recorded. One would see why in a chain—in a given fold—the location of the maximum effort moves in time, lengthwise, widthwise, or heightwise; in a word, in space. One would catch, as if in movement, the role of time required by the internal resistances of the object undergoing deformation and the role of the external obstacles: one would infer from them the reasons that the peak of the paroxysm occurring here earlier, there later. One would witness, to a certain extent, the reciprocal fight undertaken by the arcs for the possession of space; the stretching they display at their free extremities while advancing and unequally warping along the transverse alignments; the complicated behavior of nappes hatching at depth, under the misleading quietness of the carapaces; and a thousand other features that I shall leave out. My purpose shall be only to indicate for a given regional tectonics one salient feature, and for the whole I shall limit myself to a few general aspects. (205)

Here are the Himalayas: great gaps still separate the investigated areas. For a long time, however, the main feature, which is overfolding toward the south, has been understood.

The role of the crystalline zone of the high summits—the Himalayan zone—has been compared to that played in the Alps by the Mont Blanc massif. I shall attempt to sharpen this analogy. One is indeed dealing with a Precambrian, marginal fragment of peninsular India: a fragment first warped in a great basement fold, under an effort to which the energy restituted by the nappes of the Tibetan zone contributed; then broken up into sliding wedges, in clean-cut nappes and thrust southward by the continuation of the same effort. One knows that in the Alps such is indeed the usual behavior of the Hercynian massifs of the first zone, under the northward push of the Pennine nappes. In favor of this hypothesis, one could mention the occurrence of Gondwana beds in the main thrust plane and also the fact that in the extension of the Algonkian bundles of folds of the Aravalli Range, so characteristic of the northwestern flank of peninsular India, one finds in the middle of the Himalayan zone folded formations to which such an age has been attributed: similar relationships occasionally occur in the first Alpine zone between the external Hercynian massifs that have essentially remained in place and the more internal Hercynian massifs that are broken up into thrust wedges. The deposits, almost flat-lying and still little known, that were reported in the upper parts of Mount Everest most probably belong to some higher tectonic level: the normal cover of the crystalline of the Himalayan zone or the nappes originating from the north. Furthermore, recently explored traverses north of Mount Everest have shown new Nummulitic synclines that correspond fairly well to the extension of the band, known for a long time, in the High Indus. With respect to the western part of Tibet, the geologic contours that E. Hennig has inferred from the observations made, and the collecting done, by A. Sven Hedin show a great northward extent of the Jurassic and Cretaceous sediments of the Tethys: this Alpine frame, which is cut across by numerous massifs and pointings of granitoidal rocks, supports Cenozoic lavas that predominate eastward from the source of the Indus and reach, by extending along the northern bank of the Brahmaputra, the 88th meridian.

I shall now look at the southern Iranian arc. It is clear that the most frontal portions of this chain, truncated sharply by the coastline in the vicinity of Karachi, must continue beyond the Oman Sea: indeed, they reappear in Arabia in the chain of Muscat, which has to be considered an Alpine arc since it contains folded Nummulitic.

Let us briefly glance at the cover foldings of Indo-Africa. Some of them are well known as such; others have not been understood and have been interpreted as warpings between faults; still others remain to be discovered. First, here are the Jurassic and Nummulitic chainlets of the Rajasthan, the forefront of the Iranian arc (206) in the direction of peninsular India; then several still poorly understood accidents along the western margin of Iraq; then Lebanon, with its well-ordered folds; farther away

Palestine, which consists of weakly folded Cretaceous and Nummulitic, slightly affected toward the east by faults that foretell the great Jordanian fractures; still farther away folds occur in the northern part of the Sinai Peninsula; and also in lower Egypt on both sides of the top of the delta; and in Cyrenaica, where two broad Nummulitic anticlines, which outline from south of Darnah to east of Banghāzī an arc convex to the northwest, have been interpreted as the result of vertical faults separating warped blocks—it is obvious, on the contrary, that the folds are the main feature and the faults the detail. Thus one reaches in the west the large dome with a core of Triassic and Jurassic of the southeast of Tunisia, which extends over the continent south of the island of Jerba, and the weak foldings that ripple the tabular areas of southern Algeria. It is certain that appreciable tangential efforts have been transmitted deep inside the mass of Africa, either through the cover or perhaps through the basement, as shown by the foldings in the Benue basin of northern Nigeria that involve Middle Cretaceous to perhaps Late Cretaceous deposits whose relationships with the Nummulitic are not clearly established.

Let us now consider much larger aspects. For many years, geologists have interpreted the geosynclinal foldings generated in the Mediterranean as a double chain system, with its north wing the Alps, Carpathians, and Balkan thrust toward Europe, and its south wing, the Atlas, Apennines, Dinarides, and Taurides thrust southward toward Indo-Africa, taking into account the sinuosities displayed by the arcs. This previous interpretation has some good points. The extension of the south wing into the southern Iranian arc and into the Himalayan arc is unquestionable, but what about the north wing in Asia? Whether the Balkan is connected or not with the Crimea is a question of secondary importance; the weakly folded areas reported by J. Cvijic between its oriental branchlets, in the extreme east of Bulgaria, seem rather to indicate free extremities. At any rate, a chain may subdivide itself and consequently have more than one extension. Only one appears sufficiently demonstrated: it is the Istranca Dağlari, which extends from the vicinity of Yambol toward the southeast in the direction of the Black Sea. Beyond the Devonian massif of Constantinople and of the Bosporus, there are Alpine folds in the Izmit peninsula. On this transverse alignment, or close to it, the Alpine sheaf of the oriental Aegeids, which trends north-northeast from Chios and the vicinity of Smyrna—without being directly related with our problem—outlines a drawing together with the northern Anatolian arc, which is still a poorly known element of the north wing of the double chain. Farther away, the northern Iranian arc shows in the basement of the Demavend and still farther on in the Kopet-Dagh Alpine folds overturned to the north; similarly, on the northeastern slope of the Karakorum, then between that chain and the arc of Yarkand, are other Alpine folds overturned to the northeast toward the plains of the Tarim. (207) Apparently, it seems possible to trace this north wing to a place almost opposite the Himalayan arc. It is true that for long stretches the direction of overturning of these chains remains unknown, but if backward overturnings were to be reported, one should evaluate these possible facts by keeping in mind that if this direction corresponded frequently to that of the push, it might also be exactly contrary to it. For this situation to occur, the push does not have to change direction; it is sufficient that with time the point of application of the maximum effort be displaced downward. Similar considerations would also apply to the possible extension of this north wing to at least a portion of Tibet. The largest basement folds of Asia, the western and middle Kunlun, also occur immediately to the north, and the deep-seated flux that carries them flows north-northeast, as we shall point out later.

VIII. Types of Virgations

I shall now turn to questions of fundamental importance for the future. F. E. Suess has outlined the shape of a great number of virgations; he has talked about unconfined virgations or primary virgations, about confined or secondary virgations. With respect to the movements that generate these objects, the thinking of the master is limited to brief passages of such incredibly profound implications that it is almost foolhardy to pretend to be able to visualize the original process, namely, the *movement of visual images,* which is the "lightning" from which any invention in tectonics springs. When attempting to reanimate these deformations, one succumbs very rapidly to the temptation of extending them. They appeared to me, within the universal deformation, as a world of images in motion, and the consequences that are gradually going to result from our regional investigations will perhaps make my audience forgive me for having added some of my own thinking to the picture of these deformations.

If the shape alone is considered, without any kind of interpretation, one can recognize *virgations of the first type* (Figures 1 and 2) and *virgations of the second type* (Figures 3 and 4). The former and the latter may be either simple (Figures 1 and 3) or double (Figures 2 and 4). These types can by repetition, combination, or association generate more complex virgations. Today, I shall not attempt to find out if other kinds of virgations exist.

A virgation of the first type is recognized by the fact that in the central segment, which is convex, the folded elements are closely pressed against each other; whereas at the wing or at both wings they diverge in sheaves that open in the direction of the free wings; the most advanced elements are lined up, from one end to the other of the system, in a continuous and convex front.

In the virgations of the second type, the folded elements are appreciably spaced in the central segment, which is convex or straight. They are, on the other hand, closely pressed against each other at the wing or at both wings in an association of deviated extremities that have a tendency to form a straight line; to overlap each other in an echelon (208) pattern that often interrupts the continuity of the front so that the latter is occupied, at increasing distances from the central segment, by more and more internal elements. Finally, the deviated extremities tend to become thinner in the same direction.

There are two conditions that when developed within the same object may cause us to confuse a virgation of the second type with a virgation of the first type. Whenever the deviated extremities extend to the point of forming a well-ordered bundle, reestablishing along a certain length a continuous front, and when simultaneously a portion of the central segment is buried under a cover the similarity may become quite striking but not sufficiently so to confuse an eye sensitive to the subtle testimony of shapes. In general, enough features are visible for a correct diagnosis to be reached.

If the deforming forces are now considered, one can see that in virgations of the first type, the advance, which is very pronounced in the center, decreases rapidly toward the wings; that the effective tangential effort, which depends, along any alignment, on the power and on the internal and external resistances of the medium, decreases in the same direction; that this decrease results from a diminution of the power and not from an increase in the internal resistance; that the external resistance, which is never absent, even when no obstacle occurs, is only the resistance of a frontal part of the same nature and same plasticity as the medium that undergoes folding; that its distribution is equal enough to allow the formation of a continuous arcuated front; finally, that all these features indicate a folding that occurs in total freedom or almost so, a folding that works so to speak as the unhindered rumpling up of cloth and stops only with the lack of available matter or with the cessation of the power. Virgations of the first type, as long as they are not deformed as a consequence of the heterogeneity of the medium or hindered by an external obstacle that changes them to virgations of the second type, are *unconfined* or *free* virgations.

A spectator who would encompass the entire process by looking in the direction of the push could easily compare it with the advance of an offensive arc of troops against an opponent who resists weakly and continuously gives in, but who is present everywhere, and against whom one can use only a limited number of troops. The spectator would see a center at which strong beginning frontal attacks would continue throughout the action; he would also see a left and a right wing, slowly being reinforced and stretched during the operation but always late with respect to the center, where most of the power was still concentrated. In the case of a simple virgation, there would be only one wing. Without continuing these metaphors, one will keep in mind the concept of moving aspects, which certainly gives a truthful picture, and constitutes a concise representation, of the process. This concept, developed for virgations of the first type, applies also to those of the second, with modifications that relate to the force and the disposition of the opponent troops. (209)

The most beautiful virgation of the first type in the world is the one that develops almost entirely on Iranian soil, on the right or western wing of the internal bundles of the southern Iranian arc. The main push is directed southward. The alignment of maximum effort corresponds to the 63rd meridian; the greatest compression occurs on both sides of the 27th parallel. From this central segment, the sheaves spread out toward the northwest. There are two main sheaves; the external one, which in the general direction of Kerman and Yazd reaches farther on the 52nd meridian, and the internal sheaf, which stretches toward Bejestan on the 58th. Several free extremities extend northward beyond the 34th parallel. Farther away, if we use the same comparison we did before: the amount of cloth to be rumpled up becomes insufficient because of the nearness of the northern Iranian arc, which acts as an obstacle farther north; a few extremities bend westward and tend to take on the direction of that arc. The partitioning of the Iranian lands into synclinal deserts, into *kevirs,* results from all this spreading out. What I have said about the role of the external resistance must be considered here in a relative sense: it is quite clear that the Arabian massif acts in a powerful fashion west of the 60th and particularly west of the 56th meridian, but the bulk of this resistance is felt by the external bundles of the southern Iranian arc. The virgation of the internal elements discloses the freedom of movement of arcuated folds that still have a long way to go, toward the southwest, to reach the alignment required by the old obstacle.

Another sheaf originating from the Pamirs and consisting of more northern elements begins to unfurl in the region of Kabul and opens in a fanlike fashion toward the plains of the Helmand. Its southern folds extend by means of free extremities toward the desert, which they reach along the line Kandahar to Farah, not much different from the 32nd parallel. Its northern folds extend beyond the Farah-Herat line (62nd meridian); they stop within sight of the most internal folds of the great Iranian virgation, which had developed well in advance on the disputed territory and which now present an insurmountable obstacle oriented at right angles to any further extension of the Afghan virgation. The latter, in its central segment, which is very strongly contained by the Indian massif and already belongs to the great compression of the Pamirs, very early became a confined virgation; through its spread-out segment, it is still today an almost unconfined virgation.

It is well known that the eastern wing of the southern Iranian arc is divided into small-scale local arcs that are connected by concave inflections or separated by acute backward deflections.[14] Inflections and backward deflections are generated by local reinforcements of the Indian resistance, distributed by repeated segments. All the convex arcs display at one of their wings or at both, with variable degrees of intensity of compression, virgations of the second type: the Baluchistan arc at the left wing; the Sewestan arc at both wings and particularly on the left in the Sulaiman Range; the Salt Range arc at the left wing in the Hazara, alongside the Jhelum. The dominant (210) push, in these local arcs, is always toward the south-southeast, and the remainder of the deformation is due only to deviated stresses; all the left wings are stretched by longitudinal draggings that operate from north to south. The other details of these deformations will be discussed later on when I shall describe, in a more general way, this kind of virgation.

While dealing with the Iranian arc, let us sharpen our view of the movement by considering the vertical aspects. The great vacuum in which the main push is concentrated between India and Arabia is not only expressed by the convexity in ground plan but also by a great lowering of axis; to this more liberated flowing, a culmination is opposed at the two wings, which are strongly confined by the two great obstacles, a culmination that is revealed, in the northwestern Zagros as well as in the vicinity of the Pamirs, by the appearance under the Alpine bundles of fragments of the basement consisting of old ante-Alpine folds. In this overall view, I can neglect the detail of the secondary axial undulations, which, besides, is known only partially.

Virgations of the second type are *confined* virgations in the sense of F. E. Suess, that is, virgations created by an external obstacle. Are there other kinds of confined virgations besides those of my second type? I would be inclined to say no, but to be entirely certain a complete inventory of all the shapes displayed by confined virgations is necessary. My purpose being to see rather than to classify, I shall not discuss this problem for the time being. In order to generate virgations of the second type, an obstacle merely present is not sufficient: an obstacle parallel to the trend of the folds would generate only a breaking of waves of similar orientation. The obstacle should in addition display peculiar aspects, the most effective being a change of direction of the resistant margin. The spectator, oriented as in the previous case, would see each offensive arc in the center advance straight ahead, at a variable speed and to a variable distance according to the force of the obstacle; he would also see at the wing or at the wings the fight becoming exalted and concentrated in front of a powerful opponent to whom flanking operations were permitted.

In virgations of the second type, the arcuate arrangement is imposed from the outside and not, as in virgations of the first type, from within. Away from the obstacle, the flow lines of the general plastic flux are more or less parallel to each other, and the relatively gentle folds that are formed trend at right angles to these lines. This arrangement is maintained but becomes accentuated upon approaching and hitting the obstacle, although only in the central segment, where the resistance takes place with full strength perpendicularly to the flux lines; the folds become tightened along trends parallel to the resisting margin, they rise and sometimes are thrust over. The attack is clearly frontal. Whenever this segment of perpendicular incidence corresponds to a minimum of resistance, there is a tendency to form a convex front if we are dealing with a maximum of resistance. The front shows an inflection opposite a long segment and a backward deflection (211) opposite a short segment, which by the way may be reduced to an acute promontory. But, as the margin of the obstacle changes direction while the general flow direction of the flux remains unchanged, the folds generated in a central segment cannot stretch indefinitely; they encounter the resistance of the obstacle at the wing, and their extremities bend with a tendency, often carried to an extreme, to align themselves parallel to the direction of the margin. Thus is generated the arrangement in echelon in which the elements relay each other and outline a false continuous front. The breaking, when it occurs at a wing, may be carried as far as an imbrication, each local arc tending to override the extremity of the more external arc. Furthermore, the flux lines lose their parallelism when approaching the oblique obstacle: they diverge and deviate more and more from the central alignment, thus decreasing or eliminating the obliquity of the accosting. In the wing of a virgation of the second type, a dragging occurs along the resistant margin; I compared this process ten years ago,[15] as to its appearance, to the longshore current generated by the oblique breaking of waves along a shoreline. In the case of folds, this process is complicated by a longitudinal traction exerted by the central segment on the extremities, which, worked upon in the same direction by all these longitudinal stresses, stretch and thin to an increasing degree as one deals with transverse alignments that are more and more distant from the center. The length of the dragging and the intensity of the stretchings decrease from the resistant margin toward the open, toward the *upstream of the flux;* all the longitudinal effects decrease and finally die out in that direction. The repetition of an oblique resistance on the other side of the segment of perpendicular incidence generates a double virgation of the second type. This situation occurs, for instance, in the

Sewestan arc. The Burman arc, with a general push westward, displays at the right northern wing a virgation of the same type, although simple, if no attention is paid to details.

The delicate deformations of virgations of the second type being elucidated, one can appreciate the new interest related to this process. This interest allows us by a single look at the ground plan to detect at depth, in the forward areas, invisible resistances; to show the changes of direction of the resisting margin; to recognize beneath a chain or a row of chains a fundamental feature, namely, the general *direction* of flow of the deep-seated flux that carries the virgation; and to evaluate even the deviation displayed by the flux lines in the vicinity of anything that resists. A mistake in the direction of flow is impossible unless the object itself is not a virgation of the second type. The process also allows us to restitute some aspects of the shapes successively assumed by the folded elements; it is sufficient mentally to bring back these elements by increasing the radii of curvature of the ground plan and by taking care to preserve the tectonic style and by gradually shortening the folds, from the wing toward the center. Yet, one should avoid carrying the restitution too far upstream. This is how tectonics goes back into time; by going forward in time, one would see that except (212) for a variety of perturbations the curvatures of the ground plan become more pronounced by accentuating the same style and that the present condition is not a veiled but, on the contrary, a sharpened survival of previous states. This point is equally true for virgations of the first type, with the same consequences for the restitution: to shorten slowly the wings while loosening the front and bringing the latter upstream to a greater degree than the wings; to reduce the entire system to a frontal segment that is increasingly shorter, with fewer and fewer folds while everything is being lowered; and, finally, to wipe out everything and return all things to an almost flat condition that is the starting point of virgations of the first type. These virgations are little capable except in the case of rather broad geosynclines of occurring elsewhere but in cover or basement foldings. I do not think it is necessary to add that if virgations of the second type reveal the direction of flow of deep-seated fluxes, they tell us nothing about the direction of push; the latter has an appreciable chance to be the same as that of the flux. Of course, the opposite situation may also occur; it depends on the manner with which the high parts with their relative rigidity and their heterogeneity behave with respect to the plastic deformation, which works in volume at greater depth.

With respect to the nature of the elements that participate in virgations, there are virgations of autochthonous folds, which are also virgations of cover foldings; virgations of recumbent folds, mostly displayed by the deformation of the digitations and the secondary folds of the carapaces, as I have shown for the internal part of the western Alps, where one is dealing with a virgation of the second type; virgations of chains or of cordilleras; virgations of basement folds, like those of the Rocky Mountains, of the Tien Shan, and the one we shall detect, with consequences to be discussed later, in the Nan Shan, the Altyn Tagh, and their rear chains.

IX. Deformation of Old Eurasia

The Alpine deformations of the old Eurasia north of the Tethys are going to be examined now. I shall recall briefly the conditions that preceded them. The pre-Devonian foldings and the Hercynian foldings were leveled long ago. Erosion was accentuated toward the end of each orogenic cycle, although counteracted in vain by late foldings. During the first part of the next cycle, a few fault movements, a few foldings, precursors of the forthcoming new order, were able to delay the inevitable completion, although only to the extent the locations permitted. But long before the paroxysms of the current orogenic cycle, the condition of peneplain had rather generally been reached. The Angara beds, which are to be interpreted as the debris of the relief of Hercynian times, extend into the Liassic and sometimes into the middle Jurassic; therefore, during these times, reliefs to be destroyed later still occurred, but peneplaination, already extending over large surfaces, was reaching its end. (213)

During the Permian, the Mesozoic, and the Nummulitic, the northern margin of the properly geosynclinal Tethys constituted the area that was later to be the highlands of Tibet. Immediately to the north, in what was going to form the essential part of the Kunlun, between the 75th and the 105th meridian, the Hercynian and older frames were warped into weak basement folds; their uplifting also was helped by the repercussion of the precursor Alpine folds occurring in the Tethys. This complex of wrinkles was the first outline of the Kunlun.

During the Permian, the Scythian, and the Anisian, the sea was still able to penetrate, at least in places, this northern sloping bank of the Tethys. The sea displayed, over this narrow epicontinental margin of the geosyncline, very limited oscillations. And if the scattered patches, which are known for one or the other of these three stages, are sometimes located very close to the plains of eastern Turkestan or of the Kansu, at least as far as we know, they never reached these areas. The shoreline was not very far to the north and does not seem, on the average, to have moved very much from the line that today marks the northern foot of the chains.

It seems that during earlier times the incipient basement folds, which along this southern margin of the old Asiatic frameworks were especially well exposed to the Alpine forces, were at times a little accentuated so that the Tethys was even better contained. The main Jurassic and Cretaceous deposits are located behind the Yarkand arc, over which the Angara beds rest; the Nummulitic of the northeastern foot of the mountains is known only between Ying Kisha and Sandju, and perhaps it reached from the west this extremity of eastern Turkestan; to the east of the 79th meridian, no marine deposit younger than the Anisian has been found in the northern chains of the Kunlun or in the plains located immediately to the north. Therefore, in that segment a clean-cut barrier prevented large horizontal displacements of shorelines.

But in the area that is today the western half of the Tien Shan, the Alai and the eastern margins of the Turan displayed during the Mesozoic and Nummulitic preparation phase of the Alpine cycle a completely different behavior. This region belonged, then as today, to the eastern half of the broad meridian depression that involves, as previously said, all the width of Asia, from the Kara Sea to the sea of Oman. The northern part of this depression, having an easy connection with the Tethys through the south Turanian areas, often was covered by the sea during the Alpine cycle. Along its eastern margin, which included the regions just mentioned, the series are essentially epicontinental and discontinuous: at least the sea that did not encounter any complete barrier was able to reach very far northward. During the Triassic, the sea is present in the Alai and in eastern Bukhara. Some marine Jurassic, still poorly known, has been reported in the latter area. Middle and Upper Cretaceous (214) deposits reach the Fergana and the western half of the Tien Shan. The Nummulitic deposits extend very far over the Turanian regions, and it is perhaps from there that the sea spread eastward to Sandju. East of the 80th meridian neither marine Cretaceous deposits in the Tien Shan nor Nummulitic along the Kunlun are known. With the exception of local sinuosities, the trend of the eastern shoreline of the seas was from south to north.

The oriental segment of the Tethys, which curves from eastern Tibet into Burma, also spread eastward, over large portions of Szechwan, Kweichow, Yunnan, and Tonkin, an apron of epicontinental marine Triassic. In Yunnan, the Middle Triassic transgression is the major episode of this process.

In the Middle Jurassic, there were not many old lands that preserved any appreciable relief. However, there are a few along the periphery of which the series of continental deposits reaches a much higher stratigraphic position. One example is the continental Cretaceous of the Szechwan basin, at the base of which a Lower Cretaceous with Wealdian facies occurs. Obviously, the old basements at the periphery of the basin were warped by a push that reactivated erosion: therefore, the Szechwan still belongs to the eastern margin of Asia, which was so clearly affected by the deformations of the Andean subcycle. One sees the interest that a discovery of continental beds that are younger than the Jurassic layers known until today within the Angara series would present in other regions of inner Asia.

When the Cenozoic paroxysms occur, the Tethys is obstructed by chains. Peninsular India in the south and old Asia in the north are welded together. The movements become unified: from Cape Comorin to the Arctic Ocean, everything undergoes deformation. Strongly reinforced by the intense compression and somewhat increased by restituted energy—in the south by the chains of the Tethys and in the north by the Periarctic chains—intracontinental energies work upon Angara Land. All old Asia undulates, through its basements unequally indurated as well as through its unconformable covers; while undulating, it breaks in places. This, together with isostatic readaptations, is the essential part of the deformation process.

X. Cover Foldings, Basement Foldings, Reactivation

As previously said, the tangential deformations of continents include cover folds and basement folds; besides, it is conceivable that both effects may be superposed on the same vertical line.

Cover foldings can assume all shapes and display any radius of curvature: ordinary folds, with small radius, rectilinear or arcuate, isolated or associated into parallel or diverging bundles, and well ordered in their cylindrical shape as the geosynclinal chains are ordered on a much larger scale; folds with large radius and of restricted extent, which are (215) the initial forms of the preceding folds or those of domes and basins; folds with large radius and moderate extent, which are the basins and domes themselves. The occurrence of large radii reigning by themselves over very large areas implies something more than cover folding and indicates the participation of the basement even when the latter is not visible. Plastic levels that subdivide the cover into unequally deformable and superposed units act as lubricating surfaces that allow horizontal gliding and facilitate folding. Quite rarely, these glidings are associated with the decollement of the entire cover, which in this case glides directly over the old frame.

All these deformations also can take place whenever an extensive area of the old basement undergoes a shortening—even a very small one—under the cover, which is therefore compelled to wrinkle in places. This case already implies a weak participation of the basement, but because of the identity of the cover effects and also because of the insurmountable difficulty in detecting this participation when the basement is not exposed, I shall still say that there is cover folding.

Cover foldings as just described may be framed or not by salients of the old basement. The first case is known to occur frequently in western Europe. Asia also offers beautiful examples of this kind: the folds of the Angara beds over the Siberian shield and at least a portion of those that affect the Triassic, Jurassic, and Cretaceous formations of the Szechwan basin are strongly framed cover folds, as are, to a lesser degree, those of the Jura or of the Paris basin. Lebanon and the Saharan Atlas resulted from a large-scale cover folding. The north of Indo-Africa, its great and regular covers spread out all of a piece, is favorable to the accumulation of small tangential efforts in the shape of cover foldings that are weakly or not at all framed: typical examples are the small foldings of the tabular lands of Egypt, Tripolitania, Tunisia, and Algeria. The continent of Angara, with its essentially discontinuous veil of continental deposits, does not offer such a widespread field of action; the arrangement of the covers within individual basins does not allow everywhere such a free deformation.

I can neglect the case, although obviously conceivable, in which each of the old peneplained folds would be reactivated in its previous arch-bends during the new cycle: it is possible that this might lead to wrinkles of the cover where it occurred, but such individual reactivations have never been reported. Without excluding them completely, one may consider them as very rare and assume that the relative rigidity of old frames can no longer allow, except under extraordinary circumstances, deformations with small radius of curvature.

Basement folds represent an entirely different world—by their immense tonnage, which may exceed that of the largest geosynclinal chains, as well as by the vast amount of energy that their generation may consume, and by many other features (216) that distinguish them from these chains as well as from cover foldings. I have previously mentioned rather briefly some of the sources of their energy, and I shall discuss that subject again. They reach their maximum importance in Asia; they are of considerable importance in both Americas, of moderate importance in Europe, and strongly developed in Indo-Africa. They are such an essential part of large-scale tectonics that I cannot fail to dedicate a certain amount of time to them (Figure 5).

A basement fold is a fold with an average to large radius of curvature; it is developed within an old folded frame, leveled and more or less indurated, independently of any noticeable reactivation of the older folds; thus is generated within the old basement itself a recent intumescence, usually longer than wide, to which the unconformable cover, whenever present, adapts itself more or less. The older folds and the older batholiths that occur in shifting but sometimes rather large numbers in the core of a basement fold, behave in an interdependent and passive way. A basement fold can encompass the width of several older folds and impose its shape on any complex association of older structures. The older folds may cross it obliquely or perpendicularly or they may be parallel to it or even display in its interior the most varied sinuosities: they represent old inert features in a new order. When several closely associated basement folds occur, the traces of the older folds may extend from one to the other. Furthermore, there is no necessary relationship between the direction of overturning of a basement fold and that of the old dead folds that occupy its core.

Basement folds, like geosynclinal folds, may be joined into chains. These *basement chains,* like the chains originating from geosynclines, are *ordered* chains; that is, they are not very different from a cylindrical arrangement of the structures. But this should be understood only with respect to the external parts—the leveled surface of the core and warped covers—and not with respect to the core itself.

In the behavior of basement folds, one finds many rules that apply to ordinary folds, namely, those that remain compatible with the lesser degree of plasticity of the medium. A general arrangement in great arcs, virgations, segmentation into local arcs, axial plunging with a succession of high and low points in longitudinal section, dissymmetry in transverse section, stretching of free extremities—all these features depend, as in ordinary chains, on the power that deforms, the internal resistance, the conformation of available spaces and of unequally resistant obstacles that are distributed in segments. But the semi-plastic flux from which the basement folds originate being of continental dimensions and no longer limited to the width of a geosyncline, one has to take into account the heterogeneity of these old blocks. Therefore, this flux contains large and less deformable masses that influence the behavior and the distribution of basement folds and that are still obstacles in a more relative sense. (217)

The definition of the peculiarities of a basement fold should apply only to the new structural shapes. For example, it would be unfortunate with respect to the interpretation of the movements to confuse the axial plunging of a basement fold with that of one of the older and indurated folds of the inside of the core.

For basement folds as for ordinary folds, axial plungings are very often explained by the intervention, at a long or a short distance, of particularly resistant massifs and by the segmentary variations of power. An obstacle in front of a basement fold will in many cases compel the latter to develop an axial culmination, and in the segments with a lesser obstacle, an axial depression will occur; moreover, variations of power may compensate for and sometimes reverse these deformations, which nevertheless remain the rule. While waiting for Asiatic and even European examples, one may think of the culmination between the Saint Elias Mountains and the vicinity of the Gulf of California shown by the large Cenozoic basement folds that consist of Andean material; this culmination holds from the 62nd to the 32nd parallel, as long as a portion of Laurentia, visible or hidden, is resisting in front. But, as soon as the basement folds at the two extremities of the immense structure are less close to the obstacle or succeed in avoiding it, they are no longer so vigorously affected in their tangential behavior and become lower southward, along Baja California, southern Mexico, and Central America, to the extent of piercing only in a few areas from under the Cenozoic frame of the Antilles. Similarly, they become lower northward across the peninsula of Alaska on their way toward disappearance under the Aleutian arc. Furthermore, there are local but still large peculiarities of this phenomenon: for instance, the effect exerted by the Colorado Plateau, a frontal spur of Laurentia. In the north and in the east this spur was itself reworked into Laramide basement folds and, in turn, influenced through its western tip the development of Andean-Alpine basement folds. This influence is expressed not only in ground plan by the great backward deflection that around the 35th and 36th parallels affects all the bundle of chains of the Great Basin, the San Bernardino Mountains, and the Sierra Nevada—a perfect example of a large dissymmetric basement fold broken in front—it is furthermore expressed by the relative exaltation of the group of mountains that is responsible for the lack of junction between the great basement synclines the Sacramento-San Joaquin Valley and the Gulf of California. The push along this transverse alignment is essentially directed east-northeast, with local variations due to the backward inflection itself. With respect to the Coast Ranges, which in these same areas display important Alpine overturnings toward the Pacific, it is sufficient in order to explain them by means of the same general push to apply its maximum effort at a certain depth and not near the surface; thus are eliminated, as was done a few years ago in the case of the Alps, the difficulties raised about the idea of a unilateral push. (218) Let us consider this other group of mountains, which from the 41st to the 44th parallel has risen between the Sacramento Valley and Puget Sound. It is an axial culmination, which over a length of 300 kilometers uplifts a basement syncline and indicates a resistance at the front, beneath the lavas of the Columbia Plateau—themselves folded with great radii of curvature. This resistance may be related to the Recent great Idaho batholith, unless it is to be attributed to Laurentian conformations located more to the east. Although it is too early to separate all over the cordilleras of South America ordinary folds from basement folds, one can finally consider the well-marked lowering of the Andes between the 35th parallel and Cape Horn beyond the Brazilian resistance.

When there are two and no longer one single very solid massif, the compression of the chains and folds operates as in a vise. Except for compensation or inversion of the deformation due to the variations of power, axial culminations will take place in those chains that are most highly compressed between the jaws, as in the case of basement folds or ordinary folds. In the most important cases, such as that of Asia, the axial culmination, exceeding all that may be produced elsewhere by the action of unilateral factors, reaches the highest degree known at the surface of the Earth. If one collects all these American examples and all those of the Old World, which I shall sketch more completely in a few minutes, we should possess the dossier of a great court trial. In this exaltation or lowering of chains, there are no pure vertical movements that may be detected on a large scale and isolated from the folding. All other things being equal, the more compressed the folds, the more exalted. Obviously, if the explanation were to be in widely distributed pure vertical movements, these movements, in all the cases I am discussing and in a few others, should have affected simultaneously and in the same direction, on the one hand, an old massif and, on the other hand, the only segment of the chain that by its position could have been influenced tangentially by that massif, which often is located at some distance. Epeirogenic movements of this kind would furthermore have acted in such a way as to imitate in their distribution and intensity all the vertical deformations that one would expect from the tangential actions that are the most complicated and the best adapted to the arrangement of the spaces that characterize each particular case. A great deal of fidelity and great constancy in imitation would thus be required from epeirogenic movements. Should one in such instances refuse to admit, in a certain sense, the existence of real vertical movements? No, but since they cannot be distinguished from the vertical effects of folding, this means either that they do not exist or that they are distributed, on a large scale, in almost the same manner as these effects and are therefore related to them in some manner: they are always in the end vertical effects of folding. In conclusion, I can state that the culminations and depressions of axes, caught in their very movement, are only one of the vertical aspects of folding in action. (219)

Certainly there exist isostatic movements. But with respect to their possible distribution as the immediate vertical effects of folding, I shall demonstrate that in spite of initial appearances not one is known that may not be related, directly or indirectly, to deformations in which horizontal effects prevail or have prevailed. There is no isostatic movement detectable at present that does not have as cause, or as previous condition, a deformation of the type just described.

The problem of very large areas that display the effects of small vertical movements remains temporarily unsolved: I shall discuss it later.

With increasing curvature, basement folds break like ordinary folds, but with more grandiose effects since the medium is as a whole larger, less plastic, and endowed with a greater energy. What is a more or less vertical faultlet intersecting, with a few meters of throw, several hundred meters of hard beds in an ordinary fold becomes here a fracture with a large throw cutting across the entire visible mass. One can visualize the danger of interpreting these fractures as radial faults and the difficulty in proving the existence of radial dislocations in regions with basement folds. The apparent analogy of the two situations can go very far. Let us consider a basement fold or, better, a series of these objects undergoing incipient deformation. The surface of the land resembles a large swell, but too many tensions develop in this indurated medium for breaking not to occur, and on a large scale: wide or narrow voussoirs that actually are *voussoirs of basement folds* overlook adjacent areas that collapse; others subside between compartments that are hence turned into salients. They are indeed in a sense horsts and grabens, and the differentiation of these objects—a simple detail in a folding with a large radius of curvature—does not prevent this folding from continuing. New tensions occur with new faults, new horsts, and new grabens: the clean-cut dislocations of the first phase are reworked and thus the process may continue from phase to phase for a long time without the disappearance of the general shape of the basement folds. On the contrary and except for secondary effects, the general shape is gradually accentuated through time because this shape, which involves such great thicknesses of semi-plastic matter, is still superficial with respect to the plastic substratum that creates it and revives it many times through its own deformations. Furthermore, these fractures, in spite of the important throw they may display, are nevertheless surface features that cannot penetrate or survive in the substratum. The cause of so many plastic and clean-cut deformations is nothing else but a tangential effort.

This is also true for the chains, geosynclinal or not, that undergo their first cycle: the relationship between fractures and foldings is of same nature as that of the basement folds, and the differences that result from a usually greater plasticity are not essential. (220) Most of the continents that we know consist of such chains and mostly of basement folds: it is uncertain whether they display any other structures than those that can be ascribed to these two great types. Besides, one can easily visualize by considering several types of plastic deformation that a given fold may not result directly from a tangential effort, but I shall not discuss this possibility. While being folded, things break in certain places. That we certainly know very well, and the thousand nuances of deformation allow innumerable facts to be associated into a single synthetic view. In front of this harmony, in which everything merges without disappearing, where everything becomes interrelated in a completely flexible order, randomness vanishes and chaos retreats.

On this planet, there is therefore neither a fracture nor a flexure whose radial origin is today beyond dispute. It is the same for epeirogenic movements and, more generally, for any pure vertical movement. Probing deeper, one would find that the popularity of these explanations results from an economy of thought: it is much easier to visualize in one or two dimensions than in three, and furthermore, the time factor is not even considered. Perhaps the future will reveal whether these vertical deformations really operate or if they ought to be considered as simply indicating a certain deficiency in the human intuition of movement and of space.

Naturally, I do not wish to deny flatly the existence of radial dislocations or of movements with great radius of curvature independent of foldings. If the reality of these things were to be established, if their deformations were to be shown being superposed on those of foldings, then I would be delighted not to have refused to consider any of the aspects of this universe whose complexity I admire. I fully intend to use all the mechanical artifices belonging to sound tectonics, and one can see that I am adding a few more. What I wish to say is that the picture of deformation by faulting, being in essence a view of the discontinuous, can be used to construct mechanical models, to build large or small systems, but it is not sufficient for the purpose of true syntheses. One should not abandon faulting processes, but they should be put in their proper place within the continuous deformation surrounding them. Tectonics should be the visualization, then the science of deformations, before being that of dislocations: this is a necessity inherent in the process itself.

In the media involved in tectonics, there is never compression without tension. Whenever a folding occurs, regardless of its curvature, compression begins the process and continues as such uninterruptedly after the first tensions it has created have been released by fractures. In this way, it generates new fractures, and so on. Any fracture born in a compressed medium is subordinated to the deformation in the same way that the surface is subordinated to the volume. How a deformable and even rather plastic medium submitted to a general compression may undergo perhaps at the same time and in particular places tension processes is demonstrated (221) by any glacier and its crevasses without the need of any theory.

The evolution of a basement fold may remain at the stage of a more or less dissymmetrical wrinkle raised above its surroundings. When this evolution reaches the stage of overthrust, the wrinkle, which is already warped toward the front and the top, continues to be visible in the upper and frontal face of the thrusted mass. It persists while being deformed and can even display stretchings through its conflict with adjacent masses. But it preceded the overthrust. It expresses precisely the earlier character of continuous deformation with respect to clean-cut thrusting: it is in fact the *frontal swelling* of the latter, as I have defined it elsewhere.[16] Besides, this object is frequently destroyed by erosion, to which it is particularly exposed because of its protruding and advanced position. The thrusts by means of which a basement fold, usually incapable of being exaggerated into a well-developed recumbent fold, releases a portion of its internal tensions are obviously clean-cut thrusts. At present I cannot conceive that clean-cut thrusts of appreciable size could be generated at the expense of old indurated frames without a frontal swelling, that is, without a basement fold as the initial condition. The deformation may be as small as one would wish; it can become practically negligible under certain conditions of almost rigid behavior and particularly for small-scale objects that break at the topographic surface or in the subsurface; it never becomes negligible on a large scale because there is no medium that is entirely devoid of plasticity.

The style of recumbent folds is in fact excluded from these reworkings of the old frame, but such a style may develop in the cover when the latter is exposed to conditions that provide it with sufficient plasticity. Through the multiplication of thrust surfaces, the external side of a basement fold may be divided into a certain number of thrust blocks—rigid and sliding wedges that continue to displace forward their frontal swelling. When basement folds are generated along the extreme margin of a continent, adjacent to a very active geosyncline, it may happen that they are overridden by nappes arising from the latter: this is the case of the relatively small basement folds corresponding to the Hercynian massifs of the first Alpine zone and also of the much greater basement fold representing the Himalayan zone.

Basement folds being only late intumescences arisen from a foreland, they do not necessarily have, as in the case of chains rising from geosynclines, a foreland heterogeneous to them, that is, one consisting of folds much older than those forming their core. Involved in this delicate problem are at least three conceivable situations with respect to the material susceptible to deformation. The old land inside of which a basement fold arises may be tectonically heterogeneous, the oldest fragments having a rigidity higher than that of the not so old adjacent folds; (222) it may be heterogeneous, these fragments having the same average rigidity as do their surroundings; it may be tectonically homogeneous, namely, made up of folds belonging to one cycle.

In the first case, the particularly resistant margin of the very old fragments will generate basement folds at the expense of the adjacent masses of old folds, which are a little less indurated. These basement folds will have in a very real sense a *reactivated foreland*. This is the manner in which the margin of the large pre-Devonian massifs in Asia sometimes behaves with respect to the basement folds of Alpine age and of Hercynian material. It is also the manner in which the margin of the large Precambrian massifs behaves with respect to the basement folds of Alpine age and consisting of old Paleozoic folds.

In the second and third cases, the external margin of an anticlinal basement fold may well indicate, in front of a land of the same composition but less deformed, the limit temporarily reached by a particular effort; this margin also may result from the influence exerted across a certain width of old folds by a reactivated foreland that is located at a variable distance and that is particularly indurated. One shall see that this kind of influence has been exerted with great efficiency by the Siberian massif.

Thus, five kinds of behaviors exist to which may be added those that may result from secondary heterogeneities such as the presence of large batholiths surrounded by less rigid dead folds.

In summary, basement folds do not arise from geosynclines but from old basements previously folded and indurated. Entire chains with high reliefs and important thrusts are thus generated *without geosyncline,* within continental areas. Whether this happens under shallow epicontinental seas, or under fresh water, or even in dry conditions, these superficial differences do not really matter; they have no influence whatsoever either on the movement or on the essential process. *In a basement fold, the continental mass itself undergoes folding.*

The material of basement folds is less plastic, at least in the upper parts, than that of *new* chains, namely, of chains that undergo their first cycle or subcycle of deformation. The new chains, whether of geosynclinal or of Circumpacific type, include indeed a high proportion of still very plastic sediments that are completely absent in basement folds. Given the same volume, basement folds therefore require more energy than new chains. Besides, on a worldwide basis, the volume of the former is many times greater than that of the latter: the energy consumed by the basement folds is therefore a *high multiple* of that which folds new chains. Consequently, the energetic preponderance often attributed to new chains, and particularly to those arising from geosynclines, does not exist. *The energetic preponderance belongs,* with much greater dimensions, *to basement folding, which represents not only the specific reaction of continents to tangential effort but also the major expression of folding on this planet.* (223)

XI. Duel between Indo-Africa and Eurasia

The duel between Indo-Africa and what was going to be Eurasia lasted, with interruptions, from the Cambrian. It continued during the remaining

Paleozoic times, and we have observed its records as far as the Himalayan geosyncline. Just before the end of the Paleozoic, the old Eurasia was completed and formed a single mass. The Paleozoic geosynclines were filled. The Caledonian foldings had contributed to the welding of the Precambrian blocks, and the Hercynian foldings had brought the welding to completion. A certain equality was thus displayed between Eurasia and Indo-Africa, yet Eurasia remained unquestionably less homogeneous. It is under these new conditions that the duel was going to continue during the Alpine cycle, and today it is perhaps only in a quiescent phase. The fight, which was sometimes rekindled, sometimes slowed down, sometimes interrupted by withdrawals accompanied by distensional tractions, sometimes shifted by longitudinal displacements of the jaws, involved, from Assam to western Europe, a front of 12,000 kilometers, in the extension of which were located, westward, additional thousands of kilometers between Laurentia and the Brazil-Guiana massif. At the beginning of the cycle, Indo-Africa and Eurasia soon began to divide themselves into compartments; the ancient segments subdivided into shorter ones whose number increased with time and which were going to specialize into more and more particular tasks. With the passage of time, basement folds and different types of fractures appeared in the two jaws; folds became accentuated and stretched; fractures increased in number and spread out. In each segment of push, for basement folds as well as for geosynclinal chains, the partial offensives were expressed in a horizontal sense by arcs that marked at any instant the limit reached by the effort; in the vertical sense, by a determined behavior of axes, either culmination or depression, and with secondary oscillations. At both wings of an arc, or at one wing only, marginal deformations became possible; because the attacking folds were usually strongly confined, they became tightly associated toward the end of the wing in a virgation. Among all the transverse stress alignments, those located along the common boundary of two segments were characterized in ground plan by backward deflections or by inflections and vertically by particular axial behaviors. Thus, some of the major aspects of the fight are caught in motion.

During the Early Mesozoic, the new segmentation is initiated. Fractures begin to develop in the continent of Gondwana: an Indian massif is outlined with features that time will accentuate. The depression between the Ob and the sea of Oman appears and later on shall be better defined. Whether at a given moment of its history it be variably covered by the sea or dried up does not really matter; it is sufficient to notice that from then on a zone of weakness shall separate the old Eurasia into two halves. In the west is the old Europe, (224) in the east the old Asia, or Angara Land. Both of them were deeply leveled by post-Hercynian erosions. During the entire Mesozoic and the Nummulitic, old Europe will be swept by marine transgressions and regressions; Angara Land, more resistant, will display these oscillations along its margins only. This antithesis, which is well known yet not explained until now, reveals the embryonic stage of a tangential stress that is moreover of the greatest magnitude. All old Eurasia warps into a complex of extremely broad and very flattened basement folds. The stress segment, which includes Angara Land and India, was during the Cenozoic paroxysms certainly provided with an energy much greater than that which operated in the segment including old Europe on the one hand and Africa and Arabia on the other. One can see that this difference during most of the Alpine precursor times was in the same direction. The two segments displayed axial culmination, the Angara segment to a greater degree than the European one: the difference of axial elevation did not exceed a few hundred meters on the average, but it was real indeed. And one can see weakly marked since the Mesozoic one of the major features that subsequent orogenic action will accentuate. Between these two axial culminations, a saddle that is but the depression extending from the Kara Sea to the sea of Oman is located.

In addition to these extremely flattened basement folds, there were others along the southern margin of Angara Land; they were better outlined but still weak and with respect to the former slightly ahead in time. They are the basement folds whose delicate deformations we have described in the portion of the Kunlun that faces present-day peninsular India. Therefore, at a certain time during the Early Mesozoic, this great buttress began to differentiate itself from the rest of Indo-Africa, and the unconformity that in different parts of the Indian shield separates the Upper Gondwana beds from the Lower ones demonstrates this situation independently of the delicate considerations with which I am involved now. The precedence and the greater intensity of the embryonic basement folds in the southernmost segment of Angara Land facing India are most probably explainable, at least in part, by a beginning of synergy between these two objects.

The chronology of the Cenozoic paroxysms in Asia is still so little known with respect to its precise ages, variable from one place to another, that in considering the deformations, one will inevitably be led to superpose events that really occurred at slightly different moments, as through a shortening of time. But; upon consideration of the approaches used in geology, this seems to be more or less always the case whenever one deals with lengths of time and no longer with instants. Besides, this inconvenience is reduced to a certain extent in the great examples we are discussing by a relative homogeneity of behavior and of style through which the episodes thus collected (225) in the global view of a certain span of time are not necessarily incongruous. Furthermore, fine tectonic analysis, which sometimes leads—there is no need here for me to tell how—to good stratigraphy without fossils, is also capable of establishing in time, without datable deposits, phases of which we know at least the order of succession. It is often possible to tie this order through one or the other of its features to some key bed of the regular stratigraphic chronology even when the dated deposits are very scattered.

During a certain span of time included within the paroxysms, the collision of the jaws and the closing of the Tethys by the geosynclinal chains were sufficiently completed to establish between Indo-Africa and the old Eurasia a state of synergy much more perfect than that before this welding occurred. With the passage of time, this dynamic solidarity gradually increased; it is this moving flux that one has now to consider, and it is within this continuity that one should whenever possible distinguish phases and keep scores.

Let us encompass with one glance, from the plains of the Ganges to the Arctic Ocean, and from the Pacific to the Turanian depression, the extent of the old lands of Asia. Along a front of at least 2,600 kilometers, from Peshawar to the tip of Assam, peninsular India advances in a north-northeasterly or a northeasterly direction toward Angara Land. Of all the local fronts that stretch from the Pacific to Europe, this one is endowed with the greatest amount of tangential energy: therefore, the chains of the Tethys, soon dried up, raise their axial culmination in the Tibet and in the Himalayas, above anything that happens elsewhere.

Between the line that from Ying Kisha, at the southwestern corner of the Ordos, follows for 3,000 kilometers the northern foot of the arc of the Yarkand, of the Altyn Tagh, and of the Nan Shan, and the line that from the Indus to the Brahmaputra indicates the southern margin of the Himalayas is the realm of the highlands. The crushed Tethys appears in it as a median band. And although data are scanty for many Tibetan regions, one can see at least in the west the geosynclinal deposits molded into a double chain whose wings have opposite directions of overturning. The *Tibetan zone,* from which emerge nappes overturned to the south, over the crystalline of the *Himalayan zone,* is an essential element of the southern wing. The Alpine folds that between the Karakorum and the old formations of the Yarkand arc are thrust against the latter belong to the northern wing. The two margins that belong already to the jaws are molded into powerful basement folds: in the north, the western and middle Kunlun, namely, the arc of Yarkand, then the Altyn Tagh and the Nan Shan with their rear chains are thrusted northward, as we shall see; in the south, the Himalayan zone is pushed southward. It would be the only large basement fold of the margin of the Gondwana continent if during the Alpine cycle important warpings had not occurred in the Australian Alps, as well as the enormous deformations, originating from the west, that on several occasions have so strongly reactivated the old Paleozoic Andes that (226) stretch in front of the true Andes and constitute in Argentina and in Bolivia most of the eastern half of the cordilleras.

The disposition of the highlands between the plains of India and the deserts of the Tarim and of Kansu is, as far as it is known, therefore symmetrical and corresponds to the Mediterranean system, of which it is an extension. The effect of the compression is revealed not only by this symmetry but also by the exceptional importance, even for Asia, of the basement folds along both margins and particularly along the northern one. To the effects of the properly intracontinental energy are added along both margins the much more moderate effects of the energy restituted by the geosynclinal chains.

These highlands are essentially but the exaggeration of Mesozoic initial outlines: in the geosyncline, furrows displayed in that segment an axial depression; along the northern margin, during precursor times and at the beginning of the paroxysm, anticlinal wrinkles had displayed as a whole, if not in detail, an axial culmination. Unquestionably, many of these folds attempted to extend outside the segment of maximum compression: eastward, southeastward, and southward in the direction of China and Indochina; westward in the direction of the depressed zone of Iran and of Turan. This latter deformation is perhaps responsible for the emergence of the Nummulitic deposits from Sandju toward Yarkand and from Yarkand toward Ying Kisha.

XII.   Tectonic Segmentation

The products of flow and plastic deformation, which are so numerous, so complicated, so organically related and which have gradually modified the deepest parts as well as the surface of the continents and of their related blocks, become clearly organized as soon as one uses the concept of flux segments and flux transverse alignments. I wish to apply this concept to basement folds, as well as to new chains. But, from the image of a segment to the idea of a transverse alignment, there is the distance from the concrete to the abstract.

Flux segments are real volumes filled by a matter submitted to a certain flow and deformation regime. The large-scale qualitative aspects of the regime are common to the entire mass, but the quantitative details undergo gradual variations when one passes from one to the other of the transverse alignments that the segment is assumed to contain.

The segmentation displayed by the flow regime of a river under the influence of a bridge with its piers and free channels, or even better of a bed with irregularly scattered obstacles, can afford a picture similar to that of a tectonic segmentation. Upstream from the obstacles, the flux rises and illustrates axial culmination. Upstream in the free channels and downstream from them, the flux is lowered or rises less, illustrating axial depression.

One would pass from the hydrodynamic to the tectonic picture by taking into account (227) the much greater viscosity of the flux, which for the following reasons is favorable to the preservation and even to the accentuation, during a lasting deformation, of a portion of the forms successively developed: by letting intervene, with a variable degree of intensity, the much greater slowness of this kind of deformation; by changing the heterogeneity of the tectonic flux, made entirely of unequally plastic and immiscible media that influence each other in space; by visualizing the impossibility of a perfectly constant flow, and therefore of a flow regime called stationary; and, finally, by thinking of all the points that I may have omitted here because they were mentioned earlier.

It is self-evident that because of the high viscosity of the tectonic medium, the trajectory followed by a material point is much shorter than the flux lines, as long as other factors do not intervene.

The *transverse alignments* are like the *axial lines* of classical tectonics only convenient fictions created to suggest in a simple fashion the picture of a deformation in volume but not to fix this approach by inanimated linear sketches.

The transverse alignments allow us to comment on the fold regime as seen in its width, while the axial lines pertain to the longitudinal view: therein lies their usefulness. There are some peculiar transverse alignments that result from a particularly rapid variation of the plastic regime; they are generally due to unusual features displayed by the resisting massifs. Only in particular instances do transverse alignments and axial lines correspond to realities. In the concrete world, a line can occur only along the intersection of two surfaces, and a surface can occur only upon the collision of two volumes: the tectonic order cannot be an exception. The tectonic surfaces are expressed on the topographic surface by lines, but these well-known axial lines, which are well suited to direct our eyes, do in fact not direct anything. On the contrary, they represent one of the features best directed by the deformation in volume. The concept of the axial line, useful in a first approximation, has remained delicately flexible in the hands of its author [17] and of a few others. However, it has been affected by a certain bias that is typical of the human mind, or more accurately, typical of the spirit of geometry; it has too often led to rigid theoretical or graphic constructions that are completely devoid of the flexible truth so characteristic of large-scale tectonics, the fundamental element of the beauty of the world.

I would therefore be very displeased if the concept of transverse alignment were in the future to congeal into a system. I see only one case, and a rather peculiar one, in which a transverse alignment is marked by a discontinuity: it is the case of a really transverse transcurrent fault. As I proceed with the delicate survey lying ahead, I consider neither an arbitrary cutting out nor an artificial partitioning of the spaces. Nothing would be more contrary to my way of thinking than to see it emprisoned within a frame of geometric lines. In my mind, there are only moving objects whose flexibility I want to preserve completely. (228) Therefore, it should be clear that when drawing a transverse alignment through a certain number of points, I use an approximation, and when carrying this concept to its limit, I assume in general a short segment occupied by a short length of folds.

XIII. Segment of Central Asia

With the increasing progress of collision between the Indian massif and Angara Land, a segment of fundamental importance becomes outlined in the geosynclinal chains and in the old Eurasia, which for reasons of concision will be called *the segment of central Asia* (Figures 9, 10, and 12).

It is easy to locate on the east and west the present-day position of the transverse alignments that limit this segment. I say the present position because through time one should take into account not only a displacement but also a deformation of the transverse alignments.

Besides, one should be careful to state the same reservations for the location of any other transverse alignment, of any other reference line, of any concrete tectonic object anywhere on the globe.

The eastern terminator lies between the eastern tip of Assam and the elbow of the Brahmaputra to the southwest angle of the Ordos toward Kung Chang (105th meridian), and from there across the Ordos toward Kuku-Khoto (112th meridian), then along the eastern margin of the Gobi, near the Greater Khingan, while staying slightly east of Hailaerh and of Nerchinskiy Zavod. On the globe, the eastern terminator traces a curve slightly convex to the east-southeast. On a ground plan, it corresponds to the great change of direction of the Tibetan chains toward Indochina and China, as well as to the eastern extremities of the Nan Shan diverted in the vicinity of the Ordos.

The western terminator of the segment of central Asia passes through the backward deflection of the Jhelum; it trends at first northward while crossing the great compression of the Pamirs, between the 73rd and the 74th meridian, then gradually curves to the north-northeast, then more and more northeast, in the general direction of Semipalatinsk, Barnaul, and

Yeniseysk. The curve outlined on the globe is slightly convex to the west-northwest.

The opposition displayed by the direction of curvature of the two terminators is responsible for the fact that the segment of central Asia intercepts, in its middle latitudes, a length of folds of about 3,000 kilometers, as opposed to 2,600 along its southern margin, in the Himalayas, and 2,000 to 2,200 kilometers in the Siberian regions. This stretching of the middle parts of the segment is certainly due, to a great extent, to the extension forces, the longitudinal tractions that took place in the mass of the folds perpendicular to the direction of push, which was almost meridian. The matter that was strongly compressed between the Indian massif and the Siberian massif and hindered moreover by the Sinian massif and by other particularly resistant masses, displayed, besides the folding properly speaking, a tendency to flow through the escapes of the system and particularly (229) through those oriented longitudinally. This flow was in the vicinity of the eastern terminator oriented to the east, that is, toward East Asia. In the vicinity of the western terminator, it was oriented to the west, toward the Turanian segment. Other longitudinal deformations have affected various portions of central Asia, but this one seems to have involved all of it, as shown by the deformation of the transverse alignments.

The segment of central Asia, thus delimited, is the location of the maximum compressions and at the same time of the greatest axial culminations for the geosynclinal chains as well as for the basement folds that nowhere in the rest of Eurasia and of the world display so much visible tonnage.

To the east of the eastern terminator, everything is lowered, and East Asia begins in a real tectonic sense. Indeed, these countries, strongly affected by the Indian compression, react by means of several types of marginal deformations, but no longer respond to the full effect of the frontal attack of India. This also occurs west of the western terminator, where the depression extending from the Kara Sea to the sea of Oman begins. From now on we shall call this depression the *Turanian segment* because of its tectonic role, thus using the name of a part for the whole. It is obvious that East Asia, essentially submitted to the Circumpacific deformations, should not be compared in all respects to the Turanian segment, but this analogy is indicated, among other things, by the depressed situation and the marginal deformations that both regions display in regard to the segment of central Asia.

XIV. Reactivation of Eurasia during the Alpine Cycle

The continuation of the bringing together of Indo-Africa and old Eurasia begins to generate reactivations of ancient resistant massifs and of old forelands, especially in the highly heterogeneous block of the old Eurasia. Those fragments that hold on fairly well generate along their margin, through basement folds, the deformation of those parts that hold less well. With the passing of time, this kind of action spreads within the mass of the less indurated frames to areas more and more distant from the reactivated nuclei. Except for local variations of plasticity, the ex-Hercynian geosynclines are with respect to older massifs in a condition of lesser resistance so that they are reactivated as Alpine basement folds. The same is true to a certain extent for the ex-Caledonian geosynclines with respect to Precambrian massifs. Broad undulations with no clearly defined direction of overturning, simple unilateral overfoldings, or double overfoldings, all indicate these reactivations through basement folds either according to the intensity of the pressures, or the duration of the pressures, or the arrangement of the masses. We are really witnessing the folding of continents, and among the particular compressions that belong to this great phenomenon, there are many, as one can see, that are located in the ex-geosynclines.

It is clear that these reactivations, while extending themselves, generate new specializations of the segmentary regime. The new segments do not (230) abolish the deformations of the older ones but rather make them more complicated. They are shorter, and their length is dictated by that of the older massifs—or of portions of the older massifs—which intervene in the reactivation.

There is no doubt that such a specialization has gradually taken place—at different scales—in the segment of central Asia. It is also obvious that on the largest possible scale it was due to the effects of the Siberian nucleus and of the Sinian massif. It is most probable, as we shall explain later, that a third and hypothetical large, ancient massif located beneath the depression of eastern Turkestan may have participated in these deformations.

For the time being, let us draw a transverse alignment in the general direction of Kyakhta, Lhasa, and Calcutta. This transverse alignment goes through a singular point near 97° E and 40° N, where there occurs the change of the margin of the Nan Shan in the direction that is going to prevail farther west in the Altyn Tagh. It divides the segment of central Asia into two segments characterized in all respects by their flow and deformation regimes: in the west, the *Indo-Siberian segment;* in the east, the *Indo-Mongolian segment.*

Ever since this new segmentation prevailed, the strongest tangential deformations and the greatest axial culminations are concentrated in the Indo-Siberian segment. The Indo-Mongolian segment and the Turanian segment maintain through their free behavior a depressed condition in regard to the Indo-Siberian segment: while the fluxes spread better, they rise less. Such is, during the remaining portion of the paroxysm and unquestionably during the numerous Cenozoic and Quaternary replicas, the essential behavior of the huge basement folds that extend from Turan to the eastern Gobi, from the Tarim and the Kansu to Baykal, and of which the Tien Shan, the Dzungarian Alatau, the Tarbagatai, the Russian Altai, the Mongolian Altai, the Khangay, and the eastern Sayan correspond in the Indo-Siberian segment to the zone of axial culmination. The massif of the Kirgiz steppes, in the Turanian segment, rises less than the Russian Altai and the Tarbagatai. In the Indo-Mongolian segment, the Transbaykalian warpings rise less than the eastern Sayan; those of northeastern Gobi less than the Khangay and the Mongolian Altai; those of southwestern Gobi less than the Tarbagatai and the Tien Shan.

Started at the same time and considered only in a general way, such is also the deformation of the new chains and of the basement folds that form the Tibetan intumescence. Its western part, which belongs to the Indo-Siberian segment, is the narrowest; it includes not only the highest absolute elevations but also—and this is more important—the highest average elevations. Contrary characteristics are valid for the eastern portion, which belongs to the Indo-Mongolian segment; the spreading is much more pronounced; the general intumescence is smaller, and the axes, except for local rises, plunge in the direction of East Asia.

We shall encounter many other testimonies of deformations by which the Indo-Siberian segment differs from the Indo-Mongolian segment, but one can already guess (231) that the reactivation of the Siberian nucleus is largely responsible for these differences. The Siberian nucleus, with its 1,000 to 1,200 kilometers of active front, is only geometrically opposed to the western wing of the much larger Indian front. And regardless of what may be intercalated between Siberia and this western wing of India, it is obvious that the eastern part of the Indian massif faces on the average less indurated massifs, which are much less resistant than the Siberian nucleus. All that I have said about this Indo-Mongolian segment demonstrates this lesser resistance. In this very segment, the advance of the Nan Shan arcs, convex to the northeast, has a similar significance.

Does this mean that in the Indo-Siberian segment, no large and very old massif of great resistance reinforced the synergy that operated between the Siberian nucleus and the western portion of the Indian massif?

The reply to this question depends essentially on the role that may be attributed in the reactivations to the crests that are located between the Siberian nucleus and Dzungaria and to the bottom, which is hidden under Recent deposits of the depression of eastern Turkestan. This depression, called *Serindia* by Alexandrian geographers,[18] is entirely included within the Indo-Siberian segment; it occupies almost all the area between the vicinity of Kashih (Kashgar) and that of Anhsi, from the 75th to the 95th meridian. Accordingly, the Indo-Siberian segment is divided along the length of the folds into three spaces elongated from west to east. They are from south to north: the *Indo-Serindian space,* identical to the western portion of the Tibetan intumescence; the *Serindian space;* and the *Serindo-Siberian space.* The southern margin of the plains of Serindia, at the foot of the arc of Yarkand and of the middle Kunlun, indicates approximately the limit shared by the first two spaces; the northern margin of the plains, along the southern foot of the Tien Shan, plays the same role with respect to the latter two. The deep-seated frame of the Serindian space being mostly inaccessible to observation and, furthermore, almost unknown, it is obvious that for the time being the role played by this basement can be appreciated only by means of the influence it may have exerted on the deformation of the Indo-Serindian and Serindo-Siberian spaces.

XV. Serindia

The Serindian depression is almond-shaped and its length follows more or less the 40th parallel. It is rather appropriately represented by an irregular pentagon of which two sides belong to the southern perimeter and three sides to the northern perimeter. The southeastern side, from the vicinity of Anhsi to that of Keriya (Yütien), from the 96th to the 82nd meridian, is bordered by a false front consisting in reality of different chains with a variety of names—which for simplification we collectively call Altyn Tagh. The southwestern side, from the 82nd to the 75th meridian, is bordered by the arc of Yarkand. The northern perimeter is entirely bordered by the Tien Shan and its associated blocks (232). We can recognize in it a northwestern side, from the 75th to about the 81st meridian; a northern side, the shortest, from the 81st to the 86th meridian; a northeastern side, from the 86th to about the 96th meridian.

The basement folds that completely enclose the Serindian amygdaloid area display in their deformation, for each of the five sides, analogies and differences that are worthwhile pointing out. These well-exposed facts lead, on the one hand, to certain definite conclusions about the interpretation of the deformation of the Indo-Serindian and Serindo-Siberian spaces; on the other hand, these same facts lead to a reduction in scope of the numerous hypotheses regarding the interpretation of the deepest portions of the Serindian space.

These hypotheses, regardless of their nuances, can really be reduced to two major viewpoints: either the invisible deepest portions of Serindia have the same composition and the same average plasticity as the visible basement of the marginal mountains, Tien Shan and Kunlun or they have a different composition and a lower plasticity. In the latter case the existence of a Precambrian massif is almost certainly implied; that is, a massif that is homologous to the Sinian massif and concealed at depth under all of Serindia or under only a portion of it. I shall omit a third hypothesis, which would make the almond-shaped depression an extension inlier due to disrupting tractions whose maximum intensity would have been located more or less along the 82nd meridian, with a decrease toward both extremities. This third interpretation can be considered seriously only when combining it, in time, with one of the first two.

It seems clear to me that in the case in which virgations would in the marginal mountains surround Serindia, they would be—if the first hypothesis is true—virgations of the first type (unconfined); if the second hypothesis is correct, they would be virgations of the second type (confined).

The southeastern side of Serindia is bordered by a powerful virgation of the second type, which encompasses the Altyn Tagh and a large portion of its rear chains and which extends, while spreading out, toward the east, so as to occupy along the entire length of the Indo-Mongolian segment the Nan Shan and its rear chains. It is the deformation of this complex, the *middle Kunlun virgation,* that we should now attempt to understand.

What is the general direction of flow of the fluxes in this extensive area, which comprises most of the Tibetan intumescence? It should be pointed out that the overturning of the old dead folds, regardless of its direction, has no importance whatsoever when one has to evaluate that of the later developed Alpine basement folds. Patches of Angara beds pinched in the old frame of the Richthofen Mountains indicate an overturning to the northeast. This indication is very valuable, yet so localized with respect to the huge extent of the system that it is not possible to generalize its significance safely. The general convexity of the arcs of the Nan Shan in the Indo-Mongolian segment is by itself no sufficient argument. The two indications increase in value when compared; they corroborate each other but do not reach the stage of being convincing. (233)

But let us examine other interpretations that are more conclusive and of a new significance.

The change of direction of the margin of the great chains takes place, more or less, at the intersection of the 97th meridian with the 40th parallel. The transverse alignment common to the Indo-Mongolian and Indo-Siberian segments also passes through that area. This meeting is not fortuitous. It indicates for the northern front of the middle Kunlun the difference between the flow regimes characteristic of each of the two segments. If the entire mass of the system is considered, instead of just the front, these differences appear even more accentuated. East of the transverse alignment, the Nan Shan and its rear chains are characterized to a point very far to the south of the eastern Tsaidam by the near parallelism, by the rather well ordered spacing, of the folded units in which the direction east-southeast to west-northwest predominates. West of the same transverse alignment, the margin of the system, identical to the southeastern side of Serindia, is oriented west-southwest up to the 82nd meridian, in the vicinity of Keriya (Yütien); first, up to the 95th meridian one sees the deviated extremities of the northern chains of the Nan Shan, then the false front that under a diversity of names, such as Anembar-Ula, Altyn Tagh, and the Russian chain, is but the deviated extremities of the internal chains. These chains, at first parallel and aligned west-northwest as in the Indo-Mongolian segment, deviate eventually to west-southwest by conforming to the margin of Serindia, and the deviation begins at meridians that are the more to the west as the chains are the more located to the south.

Therefore, the entire system is a virgation of the second type whose central segment, the Nan Shan and its rear chains, belongs to the Indo-Mongolian segment. There the fluxes find a relatively easy flow; the Altyn Tagh, with its false front consisting of echelon structures and also with its rear chains, is the left wing, which is entirely formed of sheaves more and more compressed westward.

The deep-seated flux that carries the middle Kunlun flows, therefore, essentially to the north-northeast. The flow lines maintain this direction, or almost so, where they penetrate the Indo-Mongolian segment; but to the left, they deviate, while nearing Serindia, to the northwest by passing through the north. Thus the obliquity of the boarding is diminished with respect to the southwestern margin of Serindia. The left wing of the virgation is hence strongly confined. It is therefore clear that a considerable resistance occurs along this particular side of Serindia.

How far upstream—that is, to the south, within the large Tibetan interior—is this demand for matter toward the northeast being felt appears to be a question of detail and not of principle. The great virgation of the middle Kunlun consists essentially of basement folds, but it is possible, if the demand is still appreciable in the deep south, that it may include a few folded elements of geosynclinal origin. At any rate, it represents an important element (234) of the northern wing, displaced toward the north, of the double Mediterranean chain. It is this bundle of basement folds that emphasizes along the very southern margin of the old Eurasia the orientation of the northern, properly geosynclinal wing.

Naturally, the direction of flow of the deep-seated fluxes, disclosed by planimetry, does not prejudice the direction of overturning of a given fold in particular. This question having been discussed under its general aspect, it is superfluous to talk about it again.

The *southwestern* margin of Serindia is confronted by the arc of Yarkand. The dissymmetry acquired during the Alpine cycle by this great basement fold obviously entails a steep slope toward the northeast, facing Serindia. In the back, toward the margins of the Karakorum, Alpine folds with geosynclinal elements appear. It is the northern wing of the double chain; the arc of Yarkand itself belongs to the row of basement folds that is marginal to it. The flow lines run northeast, head-on against Serindia.

Along the entire length of the southern perimeter of Serindia, practically identical to the length of the segment of central Asia, the dominance of a flow to the northeast or to the north-northeast is clearly established. Besides, the virgation of the Kunlun is the essential argument without which the other, too much scattered indications could not be usefully coordinated today. This direction is obviously the one that was predominant, either in its purity or as a component, during the extended vicissitudes that characterized the drawing together of the Indian massif and of Angara Land.

XVI. Northern Periphery of Serindia

The Hercynian frame in which the Alpine basement folds have molded the Tien Shan has undergone during its youth at least two paroxysms: one of early Dinantian age, the other post-Dinantian. These foldings have led to the emergence of the major portion of the bundle of folds, with the exception however of a more or less complex furrow that persisted between the partially emerged folds in the north and Serindia to the south. This furrow is known over rather long distances near the northwestern margin of Serindia. It emerged only after subsequent replicas, probably of Permian age. These relationships certainly suggest the idea of a chain pushed toward the south, a chain that was first built in the north, then completed by a more external bundle of folds at the expense of a furrow that was a foredeep. This is another reason for thinking that Serindia has been a pre-Hercynian foreland for that chain.

In order to understand the deformation of the Alpine basement folds that created the present-day Tien Shan, we must visualize things first on the greatest possible scale. One then sees that the entire Tien Shan belongs to the left wing of an immense double virgation of the second type; that the Urals belong to the right wing, together with other objects that do not matter for the time being; (235) that the central segment of this virgation corresponds to the Turanian segment with the exception of Iran and that it covers western Siberia as well as the Turan.

A remarkable freedom of deformation with dominance of the flow to the south-southwest and to the south distinguishes the Turanian segment—a long corridor suitable to foldings of the largest style—from the segment of central Asia—crowded with the Indian, Serindian, and Siberian massifs—and from the more western segment—which includes the ante-Hercynian basement of the Russian platform, the Podolian massif, and the Arabian massif. This freedom of deformation is true of the young chains of Iran as well as of the basement folds generated farther north. This relative freedom is demonstrated by the almost general convexity displayed in the direction of the predominant flow by folds of all types; by the lesser compression that they undergo on the average—hence their broad spacing, their relative gentleness, their spreading in width; and by the almost perfect unity with which they keep a less exalted axial condition than that in the two adjacent segments, toward which everything rises on both sides.

The overturnings to the north that occur in the northern Iranian arc and in the Hindu-Kush originate, as in all deformations of geosynclinal chains, from secondary derivations of energy. These derivations are limited to the upper levels of the plastic flux, which is otherwise deep and powerful and involves the entire mass of the continents and expresses itself upward by means of basement folds. It is conceivable that this derivation may originate from a kind of push toward a vacuum in the direction of a preexisting Bactrian depression or from a very ancient Bactrian massif. The latter would be appropriately located on a line with the Sinian, Serindian, and Podolian massifs and would, like them, be almond-shaped and play the role of foreland of the northern wing of the double chain and of its basement folds. No matter how interesting this hypothesis might be, it does not change what I am going to say about the essential deformations of the Turanian virgation.

The central transverse alignment of the virgation, which is identical to the central transverse alignment of the Turanian segment and to the easiest flow line, trends in the general direction of Omsk, west of Akmolinsk (Tselinograd), east of Perovsk (Kzyl Orda), Bukhara; it crosses Iran close to and following the 63rd meridian. Its curvature on the globe is a little convex to the west-northwest, as in the case of the western terminator of central Asia. A certain association of common deforming causes operates similarly in the volume that comprises the two transverse alignments—let us keep this indication in mind. The direction of flow in the center changes gradually from south-southwest to south as the present-day latitude decreases. Upstream is to the north, downstream to the south. As mentioned above, the Turan and western Siberia belong to the central segment of the virgation. I have also described the general features of the basement folds that fill all that space and whose dead material is generally Hercynian with an unknown number of older fragments. The major portion is buried under coverings (236) in which Cenozoic and Quaternary deposits predominate at the surface. The warpings that the progress of analysis shall certainly reveal within these coverings will gradually explain the details of the deformation whose general aspect is indicated by anticlinal structures and the voussoirs that from place to place emerge from the steppes and deserts with a predominantly west-northwest direction: Kuchuk Tau, Sultan Hiz-Dagh, and other chainlets. The system of basement folds and of cover foldings to which these occurrences unquestionably belong, should be, with respect to the former, well ordered, although with fewer structural differences of level and greater radii of curvature on the average than at the two wings that are more compressed against old massifs.

The demand for flux downstream in the Turanian segment made itself felt far upstream in the Ob basin. But this demand has been progressively weaker and later with the increasing distance upstream: this is indeed characteristic of a high viscosity medium. Thus the massif of the Kirgiz steppes did not go beyond displaying the very large radii of curvature that usually precede a perfect cylindrical ordering. The force has been smaller and the time of deformation shorter: consequently, all is less completed than in the central and southern Turan. This is also shown by the fact that the group of culminations and depressions of basement folds that constitutes the Kirgiz massif has remained chaotic and directed by local factors; the depressed axial condition and the Turanian flow regime did not have time to develop fully, in spite of the fact that the central transverse alignment crosses the western margins of the group.

Confined by a whole series of pre-Hercynian reactivated massifs, the folds that are pushed back, deviated, and compressed against these massifs and that constitute the two wings of the Turanian virgation drag very far behind the elements of the same nature that connect or relay them across the central segment. Under the influence of these oppositions, which repeat themselves along both margins of the Turanian corridor and between which are intercalated on the side weaker zones corresponding to the ex-Paleozoic geosynclines, the flow lines take on complicated shapes that are somewhat variable over time but in which predominates a divergence, a spreading out more and more accentuated downstream, toward the margins and in the lateral sinuosities. On a smaller scale, regional or local virgations of the second type develop within the main virgation but remain subordinated to it; they tend to eliminate the obliquity of the boarding and often succeed. Thus the adaptation to the framework is completed in a tighter and tighter manner.

XVII.   Turanian Virgations

A much older Turanian virgation made of ordinary folds developed during the Hercynian cycle in the geosynclines of that age. The pre-Hercynian frame is common to this old virgation and to the virgation of Alpine basement folds that followed it and filled the same space. This (237) is why these two successive virgations display analogies resulting from the similitude of flow regimes—a similitude originating from the identity of frame—and dissimilarities that are those of basement folds with respect to ordinary folds when the degree of induration of the Hercynian frame has increased with time.

On the right margin, the common frame includes the basement of the Russian-Scandinavian platform and the probable extension of the Podolian massif, which is separated on the alignment of the Donetz basin by a weak zone; along the left margin, from upstream to downstream, the Siberian resistance, the one of the western spurs of the crest, and the Serindian resistance. The Serindian massif divided the Hercynian Tethys into two branches. From the southern branch, which extended between Serindia and the future India, arose the Hercynian Kunlun, pushed from the south against the former of these massifs; the remaining portion of the branch, farther south, gave rise to the Alpine Tethys. Compressed between Serindia to the south and the Mongolian Altai to the north, the northern branch has been completely filled by the Hercynian folding; the Alpine folding operated in this mass and uplifted from it the essential portions of the left wing of the second Turanian virgation: Tarbagatai, Dzungarian Alatau, and Tien Shan. The two branches come together at the eastern and western extremities of Serindia. In the west, both branches merged normally with the Turanian geosynclinal segment of meridian direction.

This conformation of the two margins and of their lateral anfractuosities directed the two successive virgations. I shall first examine the left wing.

The western margin of the crest tended to deviate toward the left into the Russian Altai the fluxes of Turanian origin. More local adaptations, directed by the same behavior and regulated in detail by the shape of the salient of the Alatau of Kuznetsk, of that of the Mongolian Altai, and of the intermediate reentrant, took place according to partially unraveled curves. According to W. A. Obrutchev, Hercynian directions to the north-northwest predominate opposite the Alatau; in the intermediate segment, those to the northeast; opposite the Mongolian Altai, those to the west-northwest. The flux increased in power from upstream to downstream: hence, in the north we find these broad structures among which the coal basin of Kuznetsk is a good example; and in the south, a certain dominance of more compressed structures. With respect to the Alpine basement folds, the Russian Altai displays to a much greater extent the initial stage made of broad intumescences with large radius of curvature than the cylindrical ordering that is characteristic of more advanced deformations. In that respect, the deformation of the Russian Altai recalls that of the Kirgiz massif: an effect of the upstream location. But, it shows more exaltation. This greater intensity is due to the proximity of the eastern margin, namely, the massif of the crests. In comparison with the Tarbagatai, the Dzungarian Alatau, and the Tien Shan showing well-ordered basement folds of average radius of curvature, the Russian Altai displays, on the contrary, a deficiency of deformation: this is the combined effect (238) of a more upstream location and of a greater distance from the Serindian buttress. An incipient ordering is nevertheless shown over an appreciable portion of the Russian Altai by fractures following trends between east and southeast.

From the shapeless intumescence of the Kirgiz massif, better-expressed alignments disengage themselves toward the southeast, such as the Chingiz Tau and, farther away, the Tarbagatai. The group of arcs consisting of the Saur, the Tarbagatai, the Djair, and the Dzungarian Alatau—in other words, the rear chains of the Tien Shan—is much better ordered than the depressed areas that extend to the northwest toward the center of the Turanian segment: this is because the Serindian resistance is nearer. But, in comparison with that of the Tien Shan, which alone directly confronts Serindia, the deformation of these rear chains is less energetic and less completed in all respects. The zone of culmination of the above-mentioned group of arcs extends from the 80th to the 87th meridian, and the northern promontory of Serindia, from the 81st to the 86th: therefore, it is indeed this particularly resistant and nearby salient which is responsible for this culmination. But the Tien Shan closest to Serindia is adjacent to the entire northern perimeter of the great invisible obstacle. The general exaltation is much more widespread, the tonnage much more impressive than in the rear chains. The flux generated by the compression of the Serindo-Siberian space, increased in the west by the lateral Turanian supply, ends at the margin of Serindia: it rises into magnificent basement folds, then breaks and overrides Serindia by clean-cut thrusts that are known, at least along the northwestern side of this great buttress.

Because the western spur of the Mongolian massif does not extend very far westward, it leaves the northwestern and northern margins of Serindia exposed and lets them undergo all the slow violence of the Turanian flux. This is why the participation of this flux in the general effect increases westward.

The grandiose deformation of the basement folds of the Tien Shan is generated entirely by Serindia. From the west, north of the Ili, from the northwest, between Ili and Chu, between Chu and Syr Darya, between Syr and Amu Darya, powerful branches join and rise; all of them crowd together toward the great obstacle, first in the east, then right in front of it. The compression increases from the margins of the Turanian segment up to the vicinity of the northwestern side, then of the northern side of the buttress, where the greatest resistance is located. While being compressed, the branches become exalted. This powerful left wing of virgation is like the inverted image of the virgation of the middle Kunlun; the two systems are of the same order of magnitude and almost symmetrical with respect to a point located toward the center of Serindia.

The arcs that built the rear chains display, in comparison with those of the Tien Shan, a much greater freedom of movement, such as is found in virgations of the first type. This is because these arcs, while they are close enough to Serindia to form under the influence of its northern salient, are nevertheless far enough from it to develop at ease in the middle of a Hercynian medium (239), according to their own laws. The virgation of the second type, namely, the confined virgation, is on the contrary the rule in the Tien Shan, which is compelled to adapt itself immediately to Serindia. The contrast is of the same kind as that between the internal and external bundles of folds of the southern Iranian arc.

The group of arcs that under a variety of names—Boro-Khoro, Iren-Khabirgan, Bogdo-Ula—forms the northern branch of the Tien Shan preserves some traces of this easiness, but in the most external branches the adaptation increases until it is complete along the very margin of Serindia.

The depression of Dzungaria is a bundle of basement folds minutely dislocated longitudinally in which the synclinal condition prevails. In comparison with that of the Tien Shan, this lesser Dzungarian culmination originates from the greater distance to the Serindian buttress.

The Turanian flow lines, once passed the western spur of the Mongolian Altai, display a tendency to bend toward the south-southeast in order to insinuate themselves into the western portion of the Dzungarian depression. This is why the rear chains of the Tien Shan frequently show a curving to the east-northeast on the left wing, as may be seen in the Salburty and in the Dzungarian Alatau, with its extensions in the Maili and the Djair. The folds twist themselves to remain more or less perpendicular to the flow lines, and their oblique echelon patterns tend to close the width of Dzungaria.

In front, this group of arcs arranges its assault echelons along a curved line that passes more or less through the Eli-Nor and Telli-Nor lakes to reach the river Urungu a little upstream of Bulum Tokhai. But the lateral insinuation of the Turanian flux does not stop there. The Turanian flow lines, more and more deviated from the south-southeast to the south and often slightly to the southwest, sweep the lower areas of western Dzungaria and those of the segment of the Tien Shan located downstream. This deviation is produced on one side by the northern margin of Serindia, and on the other by the fact that the flux generated by the Serindo-Altai compression, increasingly developed on the eastern transverse alignments, limits the lateral expansion of the Turanian flux.

A line convex to the east, drawn from the Altaic spur to the Serindian corner in the vicinity of Kurla (Ku erh lei), through the neighborhood of Lake Uliungur-Nor and of Urumchi (Ti hwa) with maximum convexity in the desert Dzoosotoyn Elisen near 88° E and 45° N, leaves to the west the most important effects of the Turanian influx, combined with those of the compression between Serindia and the Mongolian Altai. It also leaves to the east a space in which this compression operates with less sharing.

This eastern space, more and more deprived of Turanian input, displays new axial behaviors. The Turanian influx still participates in the rising of the Bogdo-Ula to elevations of the same order as those that prevail more to the west; but beyond this effort, it is rapidly exhausted in both transverse and longitudinal directions. This exhaustion is expressed in front (240) by the weak rising of the Pei Shan in comparison with that of both the Bogdo-Ula and the central Tien Shan. It is also expressed longitudinally by the lowering of the Bogdo-Ula in the vicinity of the 92nd meridian. But the Serindo-Altaic flux operates with increasing energy when more eastern transverse alignments are considered. Consequently, an important culmination pushes up the arc of Bar Köl and of the Karlik Tagh, which relays obliquely the Bogdo-Ula. Indeed, on this transverse alignment, the hard promontory of Irkutsk, corner of the Siberian nucleus, acts through the slightly less indurated crest. In ground plan, the maximum convexity of the Alpine basement folds that molded the crest is located on this fundamental transverse alignment. This is also the case with the culmination of the warped peneplains of the Khangay.

Favored by these circumstances, the arc of Bar Köl has vigorously grown by extending its two wings. Its right wing, formed first, prevented a subsequent lengthening toward the east of the left wing of the Bogdo-Ula. The growth of this last group of arcs had continued for a long time with a certain freedom that recalls that of the rear chains of the Tien Shan. Hence, the tendency of the Bogdo-Ula to close obliquely the Dzungarian space, a tendency much less expressed here than in the rear chains and which was terminated by the lengthening of the Bar Köl. Furthermore, from this situation also originates the lack of adaptation of the Bogdo-Ula to the direction of the Serindian margin; this accommodation was reserved to the more external chainlets of the Pei Shan.

The trough of Lukchun (Turfan), a local depression of a group of basement folds, reaches a little below sea level, but this latter fact has no importance with respect to general tectonics.

East of the 88th meridian, narrow grooves that are voussoirs of basement folds appear in eastern Dzungaria up to the desert of Narin-Khukhu-Gobi. They are aligned east-southeast and indicate the increasingly indisputable domain of the compression between Serindia and the Mongolian Altai, which is a partial aspect of the great Indo-Siberian compression. These grooves, which are numerous along the eastern transverse alignments, become increasingly rare to the west, along the margins of the major portion of the Turanian influx.

In summary, the massif of the Mongolian Altai protected eastern Dzungaria, the eastern Tien Shan, and the northeastern margin of Serindia from the major part of the Turanian influx. Facing this margin, the Serindo-Siberian compression operated by itself along great lengths, and with maximum strength along the Irkutsk-Hami transverse alignment. Facing the northern Serindian margin, this still very strong compression was added to the Turanian effort, and the sum of both effects was maximum. Facing the northwestern margin of Serindia, the further increased Turanian influx dominated the Serindo-Siberian deformation, which was, by the way, appreciably reduced.

The northern side of Serindia has generally received the frontal attack of the Tien Shan. This is not so anywhere along the northeastern and northwestern sides; therefore, the latter display local arcs and perhaps small virgations (241) of the second type that seem to blend into the main virgation.

The Pei Shan and its frontal chain, the Kuruk Tagh, run from the east parallel to the margin of Serindia as far as the 88th meridian; from there to the vicinity of Kurla, the resistant margin turns to the west-northwest and, according to certain topographic documents the branches of the bundle of folds show simultaneously a virgation of the second type; these branches become seemingly compressed into a sheaf westward, and the extremities emphasize by their echelon pattern along the Kontje Darya the new direction of the Serindian margin.

Recent remarkable works have increased our knowledge of the local arcs by means of which the Tien Shan confronts the northwestern margin of Serindia. But the axial oscillations that occur in these arcs do not originate, as sometimes thought in the past, from the intersection of two successive and orthogonal foldings. The second folding, which is assumed to trend along the meridian and which would have thus generated in the folds of a first phase a succession of high and low areas, never took place. This interpretation does not measure up to actual events; it only projects in geological times the sequence of certain analytical operations of the mind. Certainly, the Alpine deformations of the Tien Shan consisted of several phases of folding, but the effort has always been directed essentially to the south, and the reason for the local arcs, as well as for the axial behavior, must be sought—for a small part—in the segmentary distribution of the powers, the heritage of deformations located upstream; for the main part, in the segmentary distribution of the resistances downstream. One should therefore consider the strong and weak systems that are certainly repeated in an alternate pattern along the invisible margin of the Serindian massif and perhaps even, at the upper tectonic level, within the mass of the Gobi beds. The conglomerates of the Neogene basin of Switzerland were deposited in particularly large masses at the mouth of the large transverse valleys that drained the Alps after the Oligocene paroxysm. During the late tectonic phases, these masses acted as buttresses and directed the deformation of the marginal arcs, in elevation as well as in ground plan, as demonstrated by P. Arbenz. The intervention of peculiarities of the basement, combined or not with that of peculiarities of the cover of Gobi beds, in multiphase deformations similar to those just described, is the true decisive factor in these problems.

All considered, the main local arcs display, from the vicinity of Kashgar to that of Kelpin, culminations at the points of backward deflection or inward inflection and depressions at the convex salients. This is, as I have shown by many examples, the most ordinary type of axial deformation. In these areas, the same axial behavior often is common to several arcs that succeed each other from upstream to downstream, a situation that implies a well-defined segmentation of the resistance. Tendencies toward a virgation of the second type are recognizable in places by imbrications of arcs that are contiguous but of different direction. (242)

In each of the four major branches of the Tien Shan, the exalted zone extends the more toward the west as the branch is located farther to the south. On this large scale, the influence of the Serindian buttress on the axial behavior is therefore obvious, and it increases from upstream to downstream.

In the interval between Serindia and the Siberian massif, there are no objects more finely dislocated in a longitudinal way than the basement folds of the Tien Shan and of its rear chains. It is a consequence of the perfection of the cylindrical ordering among average radii of curvature.

Because of the great virgation, many portions of the Tien Shan located west of the 86th meridian have undergone longitudinal components that are directed, in the arcs, from the wing toward the central segment, that is, in general, to the west. In this manner, many arcs became pointed; numerous stretchings took place; this is the plastic aspect of the deformation. In its discontinuous form, the deformation has been facilitated by longitudinal fractures that have allowed slidings, row after row, voussoir after voussoir. On regional and local scales, these deformations have given a number of interesting features to the chain in ground plan.

While stretching toward the center of the Turanian depression, the major branches of the Tien Shan, between the Chu and the Syr Darya as well as between the Syr and the Amu Darya, bend to the northwest. In both cases, these major branches are preceded on the downstream side by more delicate ones that curve to the southwest. These agile branches indicate the tendency of the Turanian flow lines to deviate to the south, then to the southeast in the vicinity of the left margins. The major branches of impressive tonnage are slow to yield to these delicate variations of the flux; they express only its more general flow and consequently maintain the northwesterly direction which is that of the central segment of the virgation. But the light branches let themselves be dragged along in a more docile fashion. In agreement with the preceding factors, the influence of the lesser compression undergone by all the branches when entering the Turanian segment has also to be taken into account. Thus is explained the particularly divergent arrangement of the stellate pattern of the Tien Shan, repeated twice. In the southern stellate pattern, the deviation of the agile branches north of the Amu Darya already indicates the resistance of the Indian massif.

The Hindu Kush with the Afghan virgation repeats the essentials of this stellate arrangement in the geosynclinal domain, with analogies and dissimilarities in the deformation.

XVIII. Evaluation of the Serindian Question

As may be seen, the reasons for assuming the existence of a Serindian massif are plentiful. By refuting such reasons, one would be led to think that the bottom of Serindia was homogeneous with the marginal mountains, which would make that region a basement syncline. This basement syncline would be almond-shaped for having been (243) compressed at both ends and to a lesser degree in the center; it would be bordered to the north and to the south by anticlines expressing the tendency of a push toward a vacuum, or of a flow toward a depression; finally, this syncline would be slightly distorted by longitudinal deformations that would have accentuated the amygdaloidal shape in plan view. In short, this syncline would differ only in its degree of induration and in its dimensions from similar arrangements developed by the most ordinary synclines.

In favor of this hypothesis, one could put forward the group of chainlets of the Mazar Tagh, which in the vicinity of the 79th meridian stretches to the south-southeast toward the center of the desert and includes folded Paleozoic sediments. But nothing prevents us from thinking that the thrusted arcs of the Tien Shan would reach that far or that one may be dealing, although it is less probable, with a folding of the normal cover of the Serindian massif. The increasingly thick continental deposits resulting from an extended lack of external drainage cover the margin of the mountains; therefore, the central part of the Serindian massif, the part that is not covered by marginal arcs, should occupy an area somewhat smaller than that of the plains.

In the presence of so many corroborative facts, among which the best are the two enormous confined virgations of the middle Kunlun and of the Tien Shan, unexplainable without an intervening resistance, I am, in the end, compelled to assume the existence of a very old Serindian massif as demonstrated. If there were not such a massif, the configuration of the fluxes that are crowded between the Mongolian Altai and Tibet, or that spread out in the eastern half of the Turanian segment, would be entirely different—so would the pattern of the chains as well as their axial behavior.

The Serindian massif is the common buttress of the Kunlun and the Tien Shan pushed in opposite directions.

The Serindo-Siberian space, a very ancient geosyncline, has given raise in its northern half to the row of pre-Devonian chains adjacent to the Siberian massif. The southern half has, entirely or in part, subsisted as a Hercynian geosyncline, filled before the end of Carboniferous times, as shown by the behavior of the Tien Shan. In the present state of geological exploration, it is not possible to state that a margin of pre-Devonian folds, symmetrical with that of the crests, was formed near the northern margins of Serindia. It is possible that something analogous may be found in the Pei Shan or be revealed in parts of the Tien Shan by future investigations. But this margin, if it ever existed, could not because of lack of space have the same importance as the powerful chain of the crests. Therefore, the pre-Devonian deformations of the Serindo-Siberian space have been either clearly dissymmetric, with only one wing being pushed against Siberia, or of attenuated dissymmetry, with one of the wings of the double chain—the southern one—being little developed.

The Indo-Serindian space, perhaps too wide in these remote times, (244) to be designated a geosyncline, gave rise, immediately adjacent to Serindia, to pre-Devonian folds followed by Hercynian folds: this complex wrinkle was the material for the future Kunlun. The remaining part of the space to the south remained as the Alpine Tethys. The Paleozoic foldings that may have occurred along the southern margins toward India could not be compared, in terms of tonnage and intensity, with those of the Kunlun. Therefore, one finds in the two spaces that generate chains, the same kind of imperfect symmetry and of dissymmetry in the symmetry regarding the ante-Alpine deformations. The southern borders of Siberia and of Serindia were surrounded by powerful sheaves of folds. But the northern borders of Serindia and of India appear deficient in this respect, and the latter for a much longer time than the former.

In the Serindo-Siberian branch as well as in the Indo-Serindian branch of the Tethys, the geosynclinal condition was clearly expressed much longer in the south than in the north. The last furrows were filled in the first branch during the Hercynian cycle, in the second during the Alpine cycle.

All these deficiencies in the symmetry of ancient deformations, and particularly those displayed by the Indo-Serindian space, are very important with respect to the question, so much debated today, of whether the continents have a fixed position, or whether they can undergo large horizontal displacements. At least, they show that the Serindo-Siberian space was noticeably wider than it is today, and the Indo-Serindian space considerably wider than today. Indeed, in the hypothesis of fixism or of restricted mobilism, the jaws should have been almost as close as they are today, and the arrangement of the old foldings would have been more clearly symmetrical, as has happened since through increased drawing together and compression. Indeed, one can imagine that an ordinary geosyncline, which is not too wide and whose two jaws are at the same level, can generate a double chain with opposed wings; this is difficult if not impossible to visualize when the system is too wide. Therefore, it is very likely that during Paleozoic times, the center of what is called the Tethys between India and Serindia resembled more an ocean than a marine embayment and that the young Kunlun, born from a continental slope that from Serindia sloped toward the open sea, resembled by its deformation the Circumpacific chains rather than half of a double chain rising from a geosyncline.

There is no doubt that India and Angara Land have been drawn together to a great extent, that the Indian massif has undergone a *great* displacement from southwest to northeast, or Angara Land an equivalent displacement in the opposite direction, and that this amplitude results from the combination of two opposite displacements. But at present it would be a waste of time to pretend that we are able to measure these displacements by means of the unraveling of folds. In order to succeed, one would have to know not only the detail of the folded and thrusted structures compressed between India and Serindia but one would also have to carry the art of unraveling folds and withdrawing nappes to a degree (245) of quantitative precision far above that possible today. However, the order of magnitude of the movement can be estimated by other means.

The bringing closer of India, which began to play an efficient role at some point in the Alpine cycle, led to a powerful compression whose synergy encompassed the Indian, Serindian, and Siberian massifs. The fan-shaped arrangement, so characteristic and displayed today by the Indo-Serindian and Serindo-Siberian spaces, has replaced the ancient and imperfect symmetry: it is the work of the Alpine cycle.

The Indo-Serindian fan consists of new chains and of basement folds, the Serindo-Siberian fan of basement folds only. It is not necessary to recall how the overturnings of the Himalayan zone are opposed to those of the Kunlun; those of the external margin of the crest, warped with a steep slope toward Siberia, and thrusts in the same direction are similarly opposed to the overturnings of the Tien Shan. The fluxes in all these cases had a tendency to overflow the jaws, which were at the same time resistant and depressed.

According to old and new observations, the relationships between the Siberian crest and nucleus are complicated by overturnings and thrustings of the first over the sedimentary margin or over the covers of the second. It is almost unnecessary to recall the overturnings, with diverging movements to the northeast, north, and northwest, that have been reported along the contacts investigated at the periphery of the loop of the Patom. Other large horizontal displacements, attributed to folds recumbent to the northwest, have been reported west of Lake Baykal by M. Tetiaev. According to the same author, overthrusts with an amplitude of the order of twenty kilometers occur farther on, in the region southwest of Lake Baykal and of the upper part of the Angara River; they thrusted crystalline over continental Jurassic. The movements are directed toward the Siberian massif, that is, toward the interior of the amphitheater, and are supposed to be attributed to recumbent folds. Still farther away, in the Irkut valley, overthrustings of Paleozoic by crystalline, emphasized by mylonites, have been recognized by G. Frederiks; they are attributed to recumbent folds.

The intervention of great horizontal movements of the Alpine cycle must be considered as obvious in regions in which the Angara beds, and particularly the Jurassic, are embedded in the substratum. The external margin of the crest, swollen into powerful Alpine basement folds, has moved by means of thrusts that should be considered clean-cut. The existence of Alpine recumbent folds appears unlikely in a region in which there is no geosyncline of that cycle. This is true also for the transformation of basement folds into recumbent folds of great amplitude. Nowhere in the world has such a kind of transformation been observed: the low plasticity of the dead material makes the process highly improbable.

If it is true that large-scale recumbent folds occur in the crests (246) and in the regions between crest and margin, they most likely belong to one or several ante-Alpine cycles that included a geosynclinal condition.

Therefore, it would be important to distinguish by means of precise criteria recumbent folds from clean-cut thrusts, and between original clean-cut thrusts and those derived from large recumbent folds. One would thus arrive at the concept of several generations of great horizontal movements. It should also be pointed out that if the conditions reached during the Alpine cycle in these regions exclude the formation of great recumbent folds, clean-cut thrusts, on the other hand, can occur during any cycle.

A recumbent fold whose reversed limb is preserved without too much stretching will simultaneously show the overturning of the series as well as traces of the original stratigraphic gradations between beds. This is what seems to result from a comparison of the data published on the external contact of the crest at several places of the loop of the Patom, where overturnings as well as gradations between the crystalline and the Lower Paleozoic have been reported. In the present state of our knowledge, this comparison is favorable to the hypothesis of Caledonian recumbent folds pushed toward the convex periphery of the loop.

All considered, these different horizontal movements of large amplitude cannot be reconciled easily with the old hypothesis according to which the Precambrian nucleus of Serindia, buried under tabular Paleozoic, is considered the precise homologue of the crystalline complex of the crests. These horizontal movements confirm the result of my investigation into the behavior of the basement folds of the entire segment of central Asia, namely, that the Siberian nucleus is a real foreland. If such were not the case, if the basement fold from the eastern Sayan to the Patom across the Baykalian and Transbaykalian regions had involved a substratum entirely homogeneous with the Siberian nucleus, then the arrangement in ground plan, consisting of the two narrow and rapidly contorted loops of the vicinity of Irkutsk and of the region of Patom, could not have developed with that shape. The explanation of the curves with small radius could not be sought, as we have done, in the planimetric conformation and in the peculiarities of the Siberian massif forming the external resistance; it should involve only the transverse variations of power, which ought to be of an abruptness irreconcilable with the regularity of the loops. The preservation and progression of such a narrowly sinuous front, ahead of a regime of basement folds that gradually encroach upon a homogeneous medium, is, besides, inconceivable because of the high degree of induration of the matter. Therefore, the medium should not be homogeneous; in other words, there should be beneath the tabular area a Siberian nucleus whose average plasticity would be lower than that of the crystalline terrane of the crests.

Naturally, all this does not imply that Precambrian foldings (247) did not occur in the mass of the crest, and if in my working hypotheses I allowed some space for this concept, that is because until now I have had no reason to doubt the transgressive character of the Lower Paleozoic, demonstrated in certain areas, by excellent observers. I have done so with the reservations mentioned elsewhere.

The swelling of the basement folds and their transformation, at least regionally, into clean-cut thrusts reveal between the margin of the crest pushed against Siberia and that of the Tien Shan pushed against Serindia an obvious analogy with respect to Alpine deformations.

XIX. Tien Shan, Kunlun, Nan Shan

The basement folds of the Tien Shan and of the Kunlun, which Serindia separated from each other over a long distance, become closer and eventually adjacent to each other at both extremities of the massif. In the west, the chains of the Alai are compressed against those of the Pamirs. In the east, from the 96th meridian on, the long wedges of the Pei Shan follow the foot of the Nan Shan very closely and over long distances in an east-southeasterly direction. Farther east, starting at the 105th meridian, the Sinian massif divides the bundles of folds again. A southern row adjacent to this old buttress comprises the right wing of the Nan Shan arcs, strongly deviated and curved to the southeast; from about the 107th meridian on, this arrangement is relayed, along the resistant Sinian front, by a more internal branch of the Kunlun: the Chinling Shan. On the opposite margin of the buttress—the periphery of the Gobi—numerous voussoirs such as Hsi Shan, Khara Narin-Ula, Sheiten-Ula, Ta Tsing Shan, and a few others encroach upon or skirt the resistant Ordos,[19] which is slightly dislocated by basement folds and wrinkled at the surface by a few cover foldings.

The western spur of the Ordos, separator of folds, seems to be located near the place at which the Huang Ho leaves the alignment of the Nan Shan, downstream of Lanchu. A southwestern salient near Tunghsiang or slightly west of it is responsible for the relay of the Nan Shan by the Chinling Shan. A northwestern salient, overlain by the Khara Narin Ula, has generated important arc-shaped deformations.

With its great promontory of the Ordos, the Sinian massif encroaches a little on the eastern margin of the Indo-Mongolian segment. From the western tip of the Ordos, near the 105th meridian, to the eastern tip of Serindia, near the 96th meridian, the behavior of the fluxes is easy; it is the place that corresponds to the maximum advance of the Nan Shan. (248)

To the northeast of this Sino-Serindian channel and of the Nan Shan, and in the very middle of the Indo-Mongolian segment, extends a space equally favorable to spread-out deformations: the Little Gobi or southwestern Gobi. This space is framed only along its eastern margin; namely, by the Sinian margin, which stretches from the northwestern salient to the western spur of the Ordos. This margin, operating as the left bank of the Indo-Mongolian segment, has given rise to a left wing of a virgation of the second type that covers the eastern portion of that space. The western part is much more extensive and displays a greater freedom of deformation; the basement folds and their voussoirs, parallel to the front of the Nan Shan, keep in general a more depressed condition than they do at the wing—it is the central segment. Farther to the west, these objects can be followed without important changes of orientation into the Pei Shan, that is, into the external bundles of folds of the Tien Shan. Therefore, one can assume that they have been pushed like that chain in the opposite direction of the fluxes of the Kunlun; this is confirmed by the analysis of the left wing.

The transverse alignment that can be drawn in a north-northwesterly direction from the northwestern salient of the Ordos to the Galbyn-Gobi across the Khara Narin Ula shows a considerable compression due to the fact that the massif of the Mongolian Altai, which is appreciably indurated, comes very close to the corner of the Ordos. The voussoir of the Khurkhu seems to belong to the Mongolian Altai. From all the azimuths between the northwest and the southwest, voussoirs converge toward this transverse alignment while becoming compressed eastward. In a space reduced to 250 or 300 kilometers that extends from the Sinian salient to the Khurkhu are compressed not only the depressed areas that correspond to the entire interval between Serindia and the Mongolian Altai but also elements that belong to the southern margin of the latter. The culmination of the Khara Narin Ula is the vertical effect of this compression. The desert of Ala Shan, the eastern portion of the Little Gobi, represents the southern half of the converging system, that is, the wing of the virgation. Everything becomes compressed toward the margin of the Ordos, along which the Khara Narin Ula and the Hsi Shan relay each other while rising. They are, furthermore, strongly deviated with respect to the chainlets of the central segment and they also lag far behind.

It is therefore clear that the fluxes in the center were oriented to the south-southwest, against the Nan Shan. On the left wing, the Ordos tended to deviate the flow lines toward the south, then to the southeast, and in certain places to the east, as seen in the Hsi Shan.

Serindia is responsible for the pattern in opposite directions of the fluxes of the Kunlun and of the Tien Shan. The Sinian massif had the same influence along its two margins, one facing Siberia, the other the Chinling Shan. In the Sino-Serindian channel, these opposed fluxes fought for the possession of space. In front of the Nan Shan, the flux of northern origin confronted the flux of the Kunlun—a powerful eddy fed by the huge energy of the Indian massif—but lost the battle. It is not impossible that the Nan Shan did overflow its weak opponent at the surface by clean-cut thrusts, at depth by more plastic conformations. (249)

Such is the history that can be read in the subtle features of the *Ala Shan virgation*.

XX. Sino-Siberian Space

A hypothesis has been presented several times according to which a continuous Hercynian geosyncline, the extension of the one demonstrated in the Tien Shan, would stretch along strike up to the shores of the Okhotsk Sea. It is certainly plausible since in such a case the Hercynian geosyncline would have extended along the southern or southeastern portion of a Sino-Siberian space homologous to the Serindo-Siberian space. But with respect to known facts, any affirmative position would certainly be unwarranted and premature. The few known facts, already highly disputable for the Pei Shan, are even more so for the Ala Shan and the eastern Gobi. Almost nothing is known of the Greater Khingan, and concerning the problematic Devonian of the Bureya, or the demonstrated Devonian of the Patchan, we are still waiting for proof of their geosynclinal nature. As previously mentioned, we are better informed about certain portions of the long discontinuous alignment, with sometimes certain, sometimes assumed Devonian rocks, that stretches from the region of Minusinsk to that of Ayan. But this alignment, located in the middle of the crest, has a more northern position than the one that has to be extended, and the character of cover folding seems well established for the area close to Minusinsk. The *Schwagerina*-bearing limestones of the vicinity of Khabarovsk belong to the upper levels of the Permo-Carboniferous;[20] their relationships with nearby formations are not sufficiently known for us to draw any inferences. Consequently, the proposed question remains open.

It is known in more general terms how often the Upper Permo-Carboniferous is transgressive. In such a frequent case, the folds that may occur in it belong either to the late phases of the Hercynian cycle or to the Alpine cycle. In order to reach a decision, we need additional facts.

Furthermore, the uncertainty that still exists about what happens, toward the interior of Asia, to the Sinian massif often prevents our stating in precise terms a certain number of problems concerning the lands on the left bank of the Amur, northern and northwestern Manchuria, the Greater Khingan, and even a portion of the Gobi. This uncertainty affects, by the way, much more the diagnosis of ante-Alpine deformations than that of the Alpine basement folds that are so strongly displayed by their anticlinal structures, warped peneplains, longitudinal fractures, and voussoirs. While waiting for new local observations, one has to try to establish in a more precise fashion the possible location of a northwestern edge of the Sinian massif; one can do so by means of the reactivations that such an edge, if it exists, may have generated at the margin of less indurated media. But it should be recognized that the use of this criterion is somewhat restricted by the remarkable capacity of the Sinian massif itself—in the regions unquestionably belonging to it—to develop Alpine basement folds. (250)

One cannot but include in the Precambrian basement of the Sinian massif the portion of the Sikhota-Alin that seems to be the extension of the ancient formations of northern Korea and southwestern Manchuria. In the Sikhota-Alin, the Alpine basement folds, with their fractures and voussoirs, usually are oriented north-northeast. The old folds have in some places different directions, closer to east-west, which also occur in some Alpine elements.

The greatest part of the General Government of Priamur belongs to the polycyclic crest, with its semi-metamorphic sediments, its crystalline schists, and its granites. The Alpine tectonics, with its basement folds, shows in the north of the country (Uda basin) the transition from the southwesterly direction that predominates along the northernmost shores of the sea to the westerly direction that continues inland. In the central and south-central areas (lower Zeya, the mountains between Zeya and Bureya, east of Bureya), a southwest alignment predominates in the basement folds and in their voussoirs. These Alpine elements cross the Amur and display on the left bank a strong tendency to curve to the west (Ilkhuri-Alin and the Manchurian portion of the Smaller Khingan), forming with the Greater Khingan—another group of basement folds with an almost meridian direction—a system with linkages, or with backward deflections, or with gradual changes of direction, open to the south-southeast, or concave in the same direction; a system that would be generated through reactivation, by a particularly resistant spur, projected to the north-northwest, in the manner of a rather well defined wedge, in the vicinity of Tsi tsi har. The tendency toward linking seems to dominate in the Smaller Khingan as does a pattern of more gradual inflections in the Ilkhuri-Alin and in the chains that in the north separate it from the Amur; this is indeed a reactivation, with decreasing intensity with distance, that must be expected from a resistance located in the south.

Let us draw a transverse alignment passing approximately through Bodune (Fu-Yü), Tsi tsi har, the Amur at 122° longitude, east of Teptorgo and the Lena near the 114th meridian. This line displays on the globe a slight convexity to the east-northeast like the other transverse alignment, which not very far away is the terminator of the segment of central Asia. The new transverse alignment divides into two halves the great reentrant that is occupied by the loop of the old folds of the Patom region, folds that are strongly reworked into Alpine basement folds. The concavity of this loop opens in the same direction as the conformations of the north of Manchuria. Therefore, one can think that a nucleus or a spur of the Sinian massif is facing the Siberian reentrant on the same transverse alignment, and not without a certain degree of congruence. This congruence could well be, to mention it just briefly, the indication of a very old Precambrian continental displacement that could have led to the separation of the oldest nuclei of the Serindian, Sinian, and Siberian massifs (Figure 7, in three phases) and created the huge geosyncline with partially Algonkian deposits, (251) from which the crests were to rise, and a portion of which was to survive until the end of the Hercynian cycle, in the vicinity of the Tien Shan. This displacement, if it ever occurred, has been less important in the Sino-Siberian space than in the Serindo-Siberian space.

The nuclear parts of the Sinian and Siberian massifs could therefore have formed a single tectonically homogeneous mass in very remote times. Whether this was the case during all ante-Alpine times is doubtful on account of the polycyclic history of the Sinian massif and of the enormous geosynclinal accumulations in the region of the crests. A Sino-Siberian welding was completed again by the filling of this huge geosyncline, itself polycyclic. The welding became permanent only at the end of the Hercynian cycle. The fundamental heterogeneity is hence the more obvious, and while in stratigraphy one can easily assume from a certain time on the existence of a Sino-Siberian continent capable of receiving transgressive covers, the tectonician, on the other hand, must try to detect beneath the particularly synergic aspects of the movements more delicate deformations, deformations that allowed the unequal plasticity of so many odd masses to express itself until the height of the Alpine cycle.

While avoiding the hypothesis of nuclear parts more resistant than the rest of the older formations, one would implicitly admit a very unlikely situation, namely, that all these ancient formations have on a large scale the same average plasticity. The margin of each Alpine basement fold would in this homogeneous medium indicate only the limit reached by a particular effort, and the distribution of the deformations along transverse alignments would express only variations of power, variations of resistance being excluded.

In many cases the nuclear parts probably consist of particularly old Precambrian, with a very high degree of induration.

What is called the Sinian massif presumably includes several nuclei of this kind with younger and more deformable Precambrian envelopes. This heterogeneous composition would adequately explain the strong tendency of the Sinian massif to react to Alpine deformations by means of basement folds, with fractures, voussoirs, and related features. In fact, these types of deformations are strongly developed toward the two margins of the massif, Pe Chihli and southeastern Korea; they are more moderately developed, that is, they have on the average a greater radius of curvature, in the central parts of the massif, such as central and northern Korea, Liaotung, and Shantung. This situation seems to indicate that the massif, at times compressed in a transverse fashion, gave away appreciably along its two margins and less in the center. The hypothesis of strongly indurated nuclei apparently allows us, better than any other hypothesis, to understand why important rows of basement folds of great tonnage reach well into the interior of the massif, as happens particularly in the Shansi, where the basement folding has deformed continental deposits that are considered Cretaceous and are correlated with the upper beds of the Szechwan basin. (252) Numerous and sometimes intense cover foldings of Alpine age are supported by a great variety of regions of the Sinian massif.

There are many indications that the real heart of Siberia, the one that occurs, slightly deformed by Alpine forces, beneath the tabular Cambro-Silurian of the center of the amphitheater and under the very first folds of the margin, consists for its major part, or entirely, of a highly resistant nucleus. Therefore, the problem of the delimitation of a Siberian massif and of a Sinian massif, including all the Precambrian, disappears, the real problem being the delimitation of nuclear masses versus their covers consisting of younger Precambrian with foldings active during several cycles and versus the belts of Paleozoic folds. Above the ancient Siberian and Sinian nuclei, these covers—insofar as they can be of marine origin—seem to represent the epicontinental lateral equivalents of the Algonkian part of the huge geosyncline of the crests, a geosyncline that stretches in the intervening region and that has also a polycyclic history. This way of interpreting things amounts to our applying between different Precambrian cycles the usual concepts of ancient buttress, of geosynclinal foldings, of cover foldings, as if one were dealing with post-Precambrian cycles. As mentioned before, one should take into account the Hercynian cycle in the folding of the crests, and it would be dangerous, in the present state of our knowledge, to eliminate the effects of the Caledonian cycle.

It should be added that as long as the identification of the nuclear masses relies only on reactivations by Alpine basement folds and not on observed unconformities, these reinforcements of the resistance do not necessarily imply a much older basement; they could result, for instance, from the presence of a large batholith.

In the ancient formations of the upper Zeya, the Alpine deformations were in general limited to the largest radii of curvature, and the peneplain there was warped only broadly. This contrast of better-ordered objects that prevail between Zeya and Bureya, beyond Bureya, and in the extreme north of Manchuria, derives from the greater distance from the nuclear resistance that we were led to assume, at least at depth, in the vicinity of the loop of Sungari.

In summary, the best ordered among these basement folds are organized on both sides of the Amur in a bundle, generally convex to the south, that almost fills the segment between the previous transverse alignment and the Sea of Okhotsk, a pattern that seems directed by the Aldanian salient of the Siberian nucleus and by a corresponding reentrant on the Sinian side.

I shall not discuss here the influence of the nuclear conformations on the old folds. However, it should be pointed out that a backward deflection or a very acute linkage occurs around the 121st meridian between the meridian directions that are assumed to extend from the Greater Khingan across (253) the low Argun and the low Shilka and those that trend east-southeast upstream from the confluence of these two rivers. The arrangement recalls that which is displayed by the basement folds in more than one place along this same transverse alignment.

Over very great distances, the ancient directions trend to the east-southeast and to the southeast, as seen in the basins of the Zeya, of the Selemdja, of the Amgun, and elsewhere. They happen to cross almost at right angles the basement folds and their voussoirs, a situation not rare, as one knows, in Transbaykalia. In some places, connecting arcs and loops occur with changing directions.

XXI. Basement Folds of the Segment of Central Asia

The large basement folds of the segment of central Asia display in their plan, their radii of curvature, and their propensity to break a distribution and degrees that are not fortuitous.

The radius of curvature depends upon at least four factors: the intensity of the tangential effort; its duration; the degree of induration; and the distance to the margin of the nuclei capable of generating reactivations. The useful tangential effort is directed along each transverse alignment by the reciprocal behavior of the power and by external and internal resistances. With equal plasticity of the framed mass, or internal medium, the radius of curvature tends to increase with the distance to the reactivated margins. The propensity toward breaking on a large scale depends upon the degree of induration and the radius of curvature. As to the degree of induration, of resistance to plastic deformation, we shall see that it depends to a large extent—when dealing with extensive massifs in which local plasticities can be canceled out in a first approximation in favor of a mass effect—upon the unequal antiquity of the old folded basements.

Let us examine in detail the pattern of these interactions: first, the ground plan. Besides the chains with Hercynian basements, we do not know any forms that are still flexible and that recall those of chains generated from geosynclines: arcs of the Nan Shan with their sometimes minute extremities; virgations of the Kunlun or Tien Shan type; and more or less imbricated local arcs. Let us continue our examination in elevation. The Tien Shan, the Dzungarian Alatau, the Tarbagatai, and the Hercynian part of the Russian Altai all belong to the Indo-Siberian segment. The conditions of induration and of dynamic behavior being the same as a whole, it is nevertheless clear that average radii of curvature predominate in the former three chains and large radii in the latter massif, which is too far away from both the Siberian nucleus and the Serindian resistance. Besides, the Tien Shan, the Dzungarian Alatau, and the Tarbagatai display to a much higher degree than the Russian Altai great and more or less longitudinal fractures whose number and importance should obviously increase with a decreasing radius of curvature.

The Hercynian Russian Altai, the Mongolian Altai, and the Khangay in part pre-Devonian are located at the same distance from the reactivated nuclei; however, (254) what a difference in their way of breaking! Great longitudinal fractures reappear in these ancient masses, but how much more impressive than that which the Russian Altai can offer; how much steeper, as shown by the style of the valley of the Lakes, compared to what the Tien Shan displays, although the latter is in a much better position for such breaking. A rigid behavior seems also to prevail in the Transbaykalian voussoirs, where the heaviness of the plan is very striking and where basement material is, in part, very ancient. In all these ancient massifs, the deformation generally has stopped upon reaching the large radii of curvature; at this degree of basement folding, the required tensions to generate the greatest fractures were reached. In regions that are mainly Hercynian, these tensions were in general reached only after average radii of curvature were developed.

Among all these objects, the solidity of the Siberian nucleus stands out and the latter displays only the largest radii of curvature. It is surrounded by a close ring of basement folds: eastern Sayan and Transbaykalia, with fairly well ordered voussoirs. The influence of the reactivated nucleus appears here in all its strength.

In summary, the degree of induration increases with the age of the frames. The oldest massifs are those that display the least average plasticity, but this condition occurs in different degrees. How can this fact of somewhat statistical nature be explained?

The recrystallization of sediments and the intrusion of granites contribute for a while to this effect. But the first factor, by being active through repeated cycles, may lead to variations with respect to the general plasticity, variations that are not all of the same sign. The intrusion of granites, repeated by cycles, will be expressed by an increase of the resistant tonnage, hence a certain increase of the average degree of induration for the frame as a whole. The cementation, mineralization, and filling of fractures certainly contribute to the progressive induration of the frames. An important factor is also represented by the mechanical actions undergone by the rocks, actions that occur in indefinitely repeated phases during numerous cycles. These mechanical actions are expressed by intimate compressions, by *batterings* that perhaps reach their limit only very late and whose effect is comparable, up to a certain point, to that of the industrial work-hardening to which several metals are submitted. Besides, one should not adopt an exaggerated idea of these differences of average plasticity between indurated old frames. The importance of these differences pertains much more to the great sensitivity with which the basement folds react to them and to the fact that such differences are distributed throughout huge masses, rather than to the variations of the numerical coefficients that would be used to measure them.

Thus, the basement folds of central Asia contribute to the elucidation of the mechanism of basement folding in general.

The independence of basement folds with respect to the old frames themselves is visible in a hundred places. Old dead folds wind along varied arabesque patterns in the Tien Shan (255) and extend from one basement fold to another; other folds snake inside the Russian Altai and the Kirgiz massif; others penetrate the Tarbagatai in acute bevels; still others intersect the Khara Narin Ula; and there are many more examples.

XXII. The Urals

As previously mentioned, the Urals belong to the right wing of the Turanian virgation. This chain, curved by the resistance of the Russian platform, almost perpendicularly to the left wing inserted between Serindia and the Mongolian Altai, has been deprived, or almost so, of any direct compression. Set in motion solely by Turanian flow lines more and more deviated to their right, this branch displayed with respect to the left wing, during the Hercynian and Alpine cycles, deficiencies that originated from the above-mentioned unfavorable circumstances.

During the Hercynian cycle, the Tien Shan was deformed by a powerful paroxysm in the Early Dinantian and, during a subsequent paroxysm, with replicas. The Urals are not only late, since their only detectable paroxysm occurred between the Uralian and the Artinskian; they are also faulty because of the intensity of the foldings, the angular unconformity rarely being well expressed at the base of the Artinsk Sandstone.

During the Alpine cycle, the frame was deformed in a broad basement fold that was slightly dissymmetrical with a gentle Asiatic slope and a usually steep European slope: this is the effect of the reactivation incited by the Russian platform. With respect to the left wing the deficiency is first expressed by a smaller tonnage per equal length of chain. It is displayed in the axial behaviors that remained in their most incipient stages. The plateau of Ufa, which hides a rise of the buttress, compelled the Hercynian plan to bend, and this particular feature is repeated in the Alpine plan with the nuances proper to basement folds. However, an axial lowering occurs in the basement fold, opposite the plateau of Ufa. This is the reverse of the usual axial behavior, but one should not conclude that an inversion took place. Cases of this kind are explained by the weakness of deformation. As long as the tangential effort remains below a certain value, the rising of the flux behind the obstacle does not occur because it requires an excess of energy that is precisely missing. The matter flows more easily in the free segments, but since the flow, again because of lack of energy, is rather slow in such areas, the matter accumulates almost in place, forming moderately curved arcs and rising just a little. Whenever the effort exceeds the limit value, the axial rising will rapidly take place behind the obstacle, and the masses that were crowding the free segments, while becoming rapidly depleted, will display a relative lowering and will accentuate the curvature of their plan. This is the usual behavior, which is an inversion of the incipient behavior; the weak deformation changes sign while increasing in strength. (256)

From the 57th to the 50th parallel, the Hercynian plan displays in one of these segments of easy deformation a beautiful arc that is convex to the west. The folds of the margin reach the 50th parallel with a south-southeasterly direction. Farther south, the Mugodzhary starts in a south-southwesterly direction. On the 50th parallel, there is, therefore, a backward inflection of the Hercynian plan, analogous to that generated by the Ufa plateau. But the new resistant conformation required by such conditions is hidden beneath thick sedimentary covers. There again, the Alpine tectonics of basement folds displays a lesser axial exaltation, which extends, outlined by large remains of Cretaceous and Neogene deposits, from the 50th to the 52nd parallel. As in the case of the 56th to the 58th parallel, the Alpine saddle appears slightly offset to the north, with respect to the resistant spur. On the other hand, the Hercynian folds adapted themselves better to the two obstacles: they had the agility of new folds. The two Alpine examples show the laziness of the reactivated flux: an effect of the induration and of the insufficient energy, which does not allow everywhere a boarding that is perpendicular to the obstacle. The Turanian flow lines are not always sufficiently deviated to the west: this is why the normal effect of an obstacle often is located in the northeast, on the upstream side.

The Novaya Zemlya and the Timan, other Alpine basement folds with Hercynian material, are related to the Urals. The sericitic schists, overlain unconformably by the Gothlandian of the Timan, represent perhaps the continuation of the Scandinavian Caledonides along the margin of the Precambrian concealed beneath the Russian platform. Nothing prevents us from assuming that this Precambrian is, in the regions in which only post-Silurian covers are visible, not only flanked but perhaps larded by Caledonian bundles of folds that undulate within its mass.

The persistence of the Hercynian plan in the Mugodzhary as far as the 48th parallel implies the extension of the Russian platform under at least a portion of the steppes to the north of the Caspian Sea. A connection between basement folds, from the Mugodzhary to the Tien Shan, between Syr and Amu Darya, is very likely: the interval corresponds to the great Turanian axial lowering. But such a connection does not prejudge in any way the layout of the linking between Hercynian folds across the Turanian segment.

XXIII. Geguli-Ergheni Arc, Russian Platform, Caucasus

The accident of the Geguli, a basement fold slightly dislocated longitudinally in front, seems to form together with the Ergheni, across the margin with meridian folds that displays an appreciable width west of the Volga, a large arcuated basement fold still coated with its envelope of sediments slightly complicated by cover folds. The Geguli itself, from Kuybyshew to Syzran, shows a first axial culmination; another develops before Saratov by means of a broad dome in which Carboniferous, Jurassic, and Neocomian crop out. The southern plunging of this culmination is slow (257); it extends for the Chalk until Kamyshin, and for the Nummulitic still farther south. Other intumescences that in some places reveal the Carboniferous outline to the west a parallel alignment in the valleys of the Don and of the Medveditsa. In the far south, close to the depression of the Manych, the Ergheni shows a change of direction to the southeast-east as if it had aligned itself with the Manghyshlak and the folds that come from the Turan.

Although the arc of the Ergheni-Geguli simulates the plan of a branch of the Turanian virgation, it is not at all certain that it really belongs to it. It could actually have originated from a regional push operating across the two-step system formed by the Caspian depression and the Russian platform properly speaking.

The Tertiary anticlines that double the Ergheni on both sides of the Volga downstream of Krasnoarmeysk, the Permo-Mesozoic patches of the Inder, of Chapchachi, of Bolshaya Bogdo, of Malyi Bogdo, and a few others, all seem to outline as a whole a second arcuated bundle. The latter displays directions or traces that are slightly confused in some places but generally concentric with the first bundle and clearly framed inside the Caspian depression by the predominant step, which runs to the west and northwest.

The sizable undulations that are distributed from the Volga to the vicinity of the Urals, in a part of the Permo-Mesozoic covers of the space included between parallels 50 and 53 and meridians 46 and 56, and that display with local incurvations the tendency to outline a more general convexity to the north—these undulations belong to the northern wing, strongly bent, of the double arcuated arrangement just discussed.

The alignments that are displayed in tabular Russia, with predominant directions to the north or to the west-northwest as well as all the conformations of which A. P. Karpinsky has shown the elegant behaviors, either parallel to the Urals or to the Caucasus—are basement folds or voussoirs of basement folds, or cover foldings, or a combination of all these features; some of them have a particularly broad style.

In the south appears a large basement fold that is surrounded by marine deposits: it is the Caucasus. It is pushed to the south as the Tien Shan, of which it repeats, with the exception of a few details, the main pattern of deformation. From the Caspian Sea to the Strait of Kerch, this great axial culmination is exalted—because of the tangential effort and according to the usual behavior—opposite a resistant massif that is, beyond the curved Iranian and Anatolian arcs, the powerful promontory that Arabia extends northward under Syrian and Mesopotamian lands. The ancient nucleus of the Caucasus transmitted its deformation to its sedimentary cover—mainly marine—and shoved it along the southern margin. This basement fold rose from the old continental frame in a marginal position with respect to the Tethys.

The well-known thrust of Lusatia, whose movement is toward (258) the southwest, repeats in the heart of Europe and on a smaller scale the essential features of the behavior of the Tien Shan on the margin of Serindia or of the Caucasus opposite the Arabian massif. It is opposite the Precambrian massif of Bohemia that there arose from the general flux and through reactivation the basement folds whose internal tensions developed into this clean-cut thrust.

XXIV. Basement Folds of Europe

From the northern margins of the Tethys to the Baltic shield and to the Hebridian massif, the entire mass of old Europe, consisting of ancient folds, undulated in wide basement folds that were dislocated. The deformation has not been essentially different from that of the Asiatic masses, which extend in width from the Kunlun to the Siberian nucleus, from the Turan to the Arctic Ocean; yet it was much weaker. Through lack of energy, Europe is an unfinished Asia.

It may seem surprising to explain Europe through Asia. But, upon close examination, it is only natural. Asia, metropolis of basement folds, displays this kind of deformation in its strength, in its grandiose fullness, in its endless variety. Weak aspects that may occur here and there are just sufficient to shed light on such behaviors elsewhere. Europe and Asia, which are arranged in one single continental mass, have largely the same history.

It may seem ever more surprising to call into existence the horsts and the basins of central Europe and of western Europe by means of a tangential effort and to interpret them—except a few particular cases—as voussoirs of basement folds and no longer as voussoirs of a radial deformation. While pressing analogies with Asia without forgetting the differences, we shall see that the former are fundamental and the latter, points of detail.

The diagnosis of the behavior of basement folds in Europe requires numerous precautions because of the smallness of the scale, the extreme complexity of the arrangement of the dead folds, and the relative weakness of the efforts involved.

As a consequence of the induration, the basement folds require a large amount of matter to develop according to their own laws: therefore, one can imagine that their deformation, for a given stress, would be hindered to a greater extent in too small a volume rather than in a large one. The incipient conformations predominate even more so when the effort is weak. Furthermore, the smallness of the scale reduces the use that one could make of the concept of average plasticity: the effects due to local plasticities are not more important in an absolute sense than they are within a large continent, but they are more important in a relative sense.

The complexity of the arrangement of the ancient frames that are of different age and distributed by small surfaces acts in the same direction.

In a continent that consists of large-scale units, a weak effort leads to weak reactivations by basement folds. The deformation often remains incipient, but at least it can take place without many obstacles and local complications. Such is the situation of old Asia between the end of (259) the Hercynian cycle and the dawn of the great Alpine paroxysms, that is, during Mesozoic and Early Cenozoic times. Moderate reactivations took place, at least along the southern margin of Serindia. On a large scale, the synergy of the Sinian, Siberian, Serindian, and other nuclei, which were rediscovered, so to speak, through their slightly less indurated envelopes by the tangential effort, maintained with small vertical oscillations the great axial culmination responsible for an Angara Land. The excellent cylindrical ordering that predominates in the basement folds of Asia may have started in some places rather early, but its completed aspect has been achieved by the Tertiary paroxysms and their replicas. According to this point of view, Europe, endowed with less energy, is almost everywhere late; this continent is today in a condition that is intermediate between the two situations of Asia that we just reviewed. The cylindrical ordering as well as the other features of the basement folds in Europe often are weak; it is as if they were gradually emerging from forms with large radius of curvature.

While the basement folds in Europe were unable to impose themselves with the same strength as those in Asia, they nevertheless reactivated the entire frame. They left their imprint everywhere. Observing the precautions previously mentioned and taking into account the circumstances just described, one will without difficulty notice in the apparent confusion of horsts and basins the usual laws of deformation of basement folds.

The warpings of peneplains, which are excellent detectors of the Alpine basement folds, can guide us as efficiently as other elements of appreciation.

At first, one will observe a distribution of the predominant styles in zones that follow each other from the ancient southern margin of the continent—a margin facing the Tethys and today covered by the Alpine nappes—to the ancient northern massifs, such as the Baltic shield and the Hebridian-Laurentian massif.

The first zone of basement folds, which is also called the first Alpine zone of Charles Lory, but in a completely different sense, consists of small-scale and narrow basement folds of an average radius of curvature and sometimes slightly fractured longitudinally, which appear in the massifs of Aar, Gotthard, Mont Blanc, Aiguilles-Rouges, Belledonne, Grandes-Rousses, Pelvoux, and Mercantour. Their history is complicated: for several of them, one can detect successive phases of intumescence alternating with episodes of rising and sinking that express maxima and minima of the tangential effort dating from Early Liassic times. The major portion of their present curvature came into being during the last part of the Oligocene paroxysm because this curvature was imparted to the overlying nappes; it was slightly accentuated during the late replicas. The most internal of these basement folds often are complicated by clean-cut thrusts; the most external and less dislocated ones are warped by means of broad anticlincal structures that are more or less cylindrical;

in front are located much less deformed areas (260) that are transitional to the next zone. These small basement folds of the first zone are essentially due to the energy restituted to the continental margin by the great Pennine recumbent folds. The proportion of energies that are properly intracontinental is not detectable in these folds with any certainty. The most southern basement folds of the Kunlun on the margin of the ancient Tethys may have been generated in the same manner.

The second zone includes the Massif Central of France and the dome formed by the association of the Vosges and the Black Forest. It is the realm of intumescences that are broad in all directions and little ordered or not ordered at all; their major fractures are subtransversal, of general meridian direction, with a slightly curvilinear trend.

The third zone includes the major part of the Armorican massif with an incipient ordering of the large radii of curvature, in a west-northwesterly direction.

The fourth zone, consisting of Cornwall, the Ardennes, and the Rhenish schistose massif, displays a slightly better ordering, associated with a somewhat smaller radius of curvature.

The fifth zone, namely, the Teutoburg and Thuringia forests, Harz, the northeastern margins of the Bohemian massif, the horsts of Silesia and Lysa Gore with their intervening grabens, displays a new and considerable progress of the ordering. Intumescences and warped troughs are aligned, except for local changes, from northwest to southeast; the most important post-Hercynian fractures are similarly oriented; they are, with respect to the Alpine basement folds, longitudinal fractures. The average radius of curvature has decreased further. The cylindrical ordering, without reaching by any means the perfection and the magnitude of the great basement folds of the interior of Asia, certainly recalls them; the predominance of longitudinal fractures is an indication in the same direction.

From the second to the fifth zone, everything points toward a reactivation generated by the ancient northern buttresses whose effects increase gradually as the basement folds experience the closer influence of the old nuclei. In that direction, one can see the radius of curvature decreasing, the ordering becoming more perfect, the alignments straightening, and the longitudinal fractures increasing in frequency. The basement folds, because of the above-described distribution and features, display the same behavior as those in Asia: the differences are of quantitative nature only. At this stage of development, the longitudinal fractures are also sufficient to reveal the main features of the deformation, but they do not predominate as completely as those in Asia over the fractures of random orientation that are formed before the ordering is well under way. This last circumstance is mainly responsible for the complication of the shape of the European horsts.

From the second to the fifth zone, the dead material is almost exclusively Hercynian; therefore, one can speak, on a large scale, of an average plasticity, which by the way allows comparisons between deformations. Truly, one could argue about the slightly greater resistance induced by the abundance of crystalline rocks in the (261) exposed massifs of the second zone and thus partly explain the deficiencies in the deformation of the basement folds in these areas. But this interpretation, even carried to the extreme, does not affect the rest of the progression.

The Precambrian nucleus of Bohemia, slightly more indurated than its Paleozoic envelopes but too small to offer the same resistance as the great Precambrian massifs of the north, has been involved in the deformation of the fifth zone of basement folds. However, its role has not been entirely passive: it did generate the Lusatia thrust and, in the same manner, a few other structures. It also maintained locally, amidst the narrow voussoirs of basement folds that predominate in Hercynian lands, a much broader and more rigid style, which expresses a different kind of reaction to Alpine efforts. With due consideration of sizes and differences, we can say that the tectonic position and role of the Precambrian nucleus of Bohemia, between the Baltic massif and the ancient southern continental margin, recalls the Serindian massif, located between the Siberian nucleus and the continuation of the same margin. Like Serindia, the Bohemian nucleus acted as a divider with respect to Paleozoic folds; like Serindia with respect to Siberia, it increased, along the transverse alignments it intersects, the synergy with the Baltic massif and the compression of the folds of any date later than its own; like Serindia, it generated reactivations by basement folds with clean-cut thrusts. Serindia lacked only the fact of being partially covered, as the Bohemian nucleus was, by nappes rising out of the Alpine geosyncline. Furthermore, it suffered appreciably along its limits from being enclosed between the Alpine basement folds that surged all along its margins and often encroached upon its borders, exalting the periphery either in the Paleozoic basement or in the Precambrian basement. The continuation of the massif to the southeast is concealed beneath the Neogene deposits of the foredeep and under the Carpathian nappes.

The visible part of the fifth zone sinks rapidly, up front, under the Oligocene, Neogene, and Quaternary filling of the large trough of northern Germany. Deep drillings have disclosed in this scarp, as well as farther north, the persistence of the structure in aligned voussoirs. The scarp itself, considered as a whole, is only the *forward portion* of a basement chain that is steeply sloped, dissymmetric, facing northeast, and particularly fractured longitudinally because of the diminution of the radius of curvature.

The trough itself forms the sixth zone, and little could be said if some of its deep-seated deformations did not become apparent, all the way to the north, in Denmark under the gently curved aureoles of the Chalk and of the Tertiary. These aureoles are the traces of folds of weak elevation but of very large radius of curvature and covering extensive areas. Thus, the predominance of very large radii over large surfaces requires the deformation of a very great tonnage, so great, as previously said, that no cover folding can account for it. Therefore, one has to assume here as probable (262) the participation of the basement of old folds, warped with a large radius of curvature and transmitting its conformation to the cover. In summary, one should assume basement folds in which a particularly broad style predominates. This does not at all imply that true superficial folds did not occur in places. The alignment of these basement folds is southeast-east, judging from the very large syncline whose axis crosses Jutland, reaches the coast between Fredericia and Aarhus, and furthermore intersects the Sjaelland while keeping the same general direction. Naturally, one is dealing here with the direction of the folds and not with the sinuosities displayed by their traces.

At present, nothing is known about the age and the nature of this reactivated basement. But the large radius of curvature indicates a strongly indurated substratum, probably pre-Hercynian. Indeed, if this substratum were Hercynian, the increase of the radius of curvature, in regard to what it is in more southern zones, would most probably correspond to a weakening of the reactivation effort. This situation is rather unlikely when approaching the buttress, except under very peculiar circumstances. Consequently, the substratum is either already the Baltic shield itself, or the large Caledonian branch that many reasons lead us to believe bends either beneath the North Sea or under the plains, southwest and south of the shield, or both.

The seventh zone, which comprises the voussoirs of Scania and of Bornholm, has encroached upon the Precambrian massif. The degree of induration is very high; only large radii of curvature are displayed, and the tensions that are necessary for the generation of longitudinal fractures were reached at this degree of basement folding, which furthermore required an appreciable release of energy in such a medium.

Farther north, one is allowed, in a certain sense, to consider as an eighth zone the major part of southern Sweden. It is an advanced and particularly exposed spur of the Baltic shield, strongly affected in that region by broad warpings and by powerful curvilinear fractures, also strongly dislocated in broad and frequently tilted voussoirs of variable behavior. In such a rigid medium, well-ordered foldings are out of the question. The great fractures must have followed very closely the incipient phase of the deformation by warping. Nevertheless, such a behavior may be simply considered an extreme case of the more flexible deformations displayed by the more southern zones of old Europe. Since it occurred with such a powerful energy precisely at the place most exposed to the tangential efforts coming from the south, these two arguments corroborate one another. Consequently, without pretending to eliminate partly the explanations by means of tension or stretching, it would nevertheless be unrealistic to think that the horizontal compressions, which have so appreciably deformed the rest of the old Europe, would have spared this corner of the Baltic shield. The most dislocated zone grades northward into regions less fractured by the Alpine deformations. The system of voussoirs in which the great lakes are located (263) still belongs to it. A line convex to the northwest that runs from Göteborg to Stockholm indicates approximately the margins of the most dislocated region.

Farther away and over the remaining part of Fennoscandia, more regularly warped areas occur. Whatever credit may be extended for late glacial and Recent times to the beautiful hypothesis of isostatic movements; whatever part may still be attributed to isostasy during more remote geological times in the explanation of more ancient and repeated warpings; whatever precautions should be observed for movements of such an extension, of such a radius of curvature, and of such a tonnage in the case of explanations by means of folding, it is nevertheless clear that basement folds represent a valuable transition between ordinary folds and deformation of this kind. Consequently, it would be dangerous in this new approach to reject the part played by the horizontal effort in such deformations. Whether the larger shields, of the type of Fennoscandia or Laurentia, for instance, might be considered as very broad basement folds, as *basement brachyanticlines,* should not be rejected at first glance, and I shall return later to this important question.

The Caledonian chain of Scandinavia is in its present state a fragment of a powerful Alpine basement fold, the effect of a reactivation with a large radius of curvature generated by the western margin of the Baltic shield. Indeed, nobody would admit that its present relief and the elevation of its warped peneplains could be attributed to Caledonian movements.

It is important to add that none of these great deformations of Alpine age, starting with the second zone, can be attributed to a long-distance repercussion of the folding of the Alps and of the other geosynclinal chains of that cycle. Just the opposite is true. The basement folds of Europe show, as I have demonstrated, a progression of the deformation, which grows with increasing distance from the Alps and decreasing distance from the northern buttresses. In Europe as in Asia, the tonnage of the basement folds is a high multiple of that of the geosynclinal chains; it consists of material that requires, given an equal deformed volume, a much greater amount of energy. Therefore, the deformation by basement folds, in fact, the folding of the entire continental mass, is by far the predominant process; the rest is only derivation of energy. The small basement folds of the first zone are in fact the only ones due to restituted energy.

The flux expressed by the basement folds of central Europe was, therefore, in general flowing toward the northeast. This general direction seems to have shown, over time, very slow changes of azimuths, usually maintained between the northeast and the north. Temporary reversals of direction probably occurred. Inflections of more local character, which can be left out here, were expressed by the flow lines in the vicinity of resistant nuclei. (264)

The well-known inflection displayed by the Hercynian plan approximately along the meridian of Paris, from the Massif Central to the Franco-Belgian coal basin, seems, after all, to have been directed by the large salient displayed by the Baltic shield in southern Norway. It is probable that the same conformation impressed itself earlier on the Caledonian bundles of folds that trend between this salient and the northern margin of the Hercynian range. While this Caledonian belt was the immediate obstacle for the Hercynian front, the Norwegian salient was the intermediate obstacle.

But let us now consider the axial behavior of the Alpine basement folds. These conformations do not have a necessary relationship with ancient trends because the direction of the flux varied through time.

Thus we can draw a transverse alignment passing by the estuary of the Gironde, the Poitou threshold, Paris, and the North Sea. This transverse alignment indicates with respect to the basement folds the axis of a depressed segment, of a zone of depression of the axes, comparable in many ways with the Turanian segment.

On both sides of this *Parisian segment,* everything is exalted opposite the Baltic massif and the Hebridian-Laurentian massif. In the east, all the basement folds, all the voussoirs, Ardennes, Sauerland, Teutoburg Forest, all display axial rising; and the whole group of horsts that extends from there as far as the Vistula indeed forms, on a large scale and exactly opposite the Baltic shield, an imperfect axial culmination, slightly chaotic, complicated by local lowerings, incompleted and reworked, but certainly obvious. West of the Parisian segment, there is an axial culmination, sometimes slight, sometimes strong: Cornwall, the Armorican massif, and the Spanish Meseta. The system considered as a whole displays the predominance of ordinary behavior: culmination opposite high or strong obstacles; lesser rising in the free segments that face low or weak obstacles (Figure 11).

In the presence of these associations, too numerous to be random, one cannot, in my opinion, refuse to admit for Europe the reality of basement folds and the tangential actions they involve.

If such a reality is accepted, how can one completely reject the idea that great meridian fractures such as those of the Massif Central, of the Rhine graben, even of southern Sweden could be partly due to tensions of folding origin, to *basement transcurrent faults?* This hypothesis may be surprising, but in fact it is less so than that of radial fractures, that is, of originally vertical deformation. Besides, this hypothesis relates very well to the one of tensional fractures by horizontal traction. Nothing prevents us from thinking that both types of deformation did alternate, and that both were capable, furthermore, of becoming complicated by vertical displacements contemporaneous with the main movement or subsequent to it, in infinitely varied combinations. In this kind of problem, the most complicated solutions often happen to be correct.

The Pyrenees, which are aligned with the semi-rigidness of a small Tien Shan (265) and almost recall the Caucasus, consist essentially of a bundle of basement folds of Hercynian material that have undergone a phase of good ordering in deforming their covers; whereas, at the same time, a furrow in front was outlined just sufficiently to look somewhat more like a geosyncline. While being accentuated, the basement folds were dislocated into clean-cut thrusts that shoved and folded the deposits of the furrow, partially overlapped them, and also generated, farther in front, some cover foldings. A few Andean deformations during Early Cretaceous and Cenomanian times preceded the main Cenozoic movements by a long while.

Among all the basement folds that were active north of the northern margin of the Tethys, within the mass of old Europe, the Pyrenees seem thus to display the most complete deformation. Since one would look in vain north of the chain for a pre-Hercynian massif capable of generating such a feature, it seems reasonable to attribute this reactivation to the most widely distributed differences of plasticity of the Hercynian frame itself.

During Hercynian times, a strong axial culmination was displayed where the Massif Central of France was going to be. Consequently, post-Hercynian erosions have destroyed, except for a few patches, the mantle of Paleozoic sediments that remained in a normal position; they have, furthermore, deeply penetrated into the crystalline terranes. What remains of the granite and of its metamorphic envelopes is at the right elevation to generate reactivations in frames of the same age and of slightly less than average induration. Besides, it is obvious that the extent of the basements, which are indurated over large areas, is appreciably greater than that of the present-day Massif Central. Therefore, one can demonstrate here the case of a nuclear mass that is of the same age as the frame in which it generates a deformation by basement folds. By this I do not want to state that more local causes, proper to the Pyrenean basement itself, may not have played a role in the location of this deformation. The fact that the reactivation occurred only along the southern side of the nuclear mass confirms, furthermore, the flow of the deep-seated flux toward the north-northeast, a direction that fits well within the most frequent azimuths under old Europe.

The entire Spanish Meseta was undulated by broad Alpine basement folds oriented from west-southwest to east-northeast without paying the least attention to the direction of the old folds but parallel to the Betic cordillera. The very flattened basement synclines contain Neogene and Quaternary deposits. They extend from the Portuguese basins to the Castilian basins, but an axial culmination, complicated by a few secondary warpings, trends in the general direction of Seville to La Coruña and prevents their deposits from building continuous furrows from one depressed area to the other. However, in the interior of the Meseta, alignments of small basins indicate the trends of the basement synclines. There is a great major syncline: Lisbon-Madrid, and the Tejo valley has been produced essentially by epigenesis from this basement syncline. Another syncline directs, with similar consequences, the valley (266) of the Guadiana from Badajoz upstream. A third syncline is visible in the direction Coimbra-Salamanca. The two Castilian basins are only the eastern depression of all these basement folds. Fractures, oriented as the basement folds, limit, over appreciable lengths in the warped anticlinal structures, the magnificent voussoirs that extend eastward, notching or separating the Castilian basins. But fractures, regardless of their throw, are still very small features within such a large deformed volume, and the energy consumed in producing them is only a minute fraction derived from the energy of deformation.

These powerful Spanish basement folds are essentially due to the intracontinental energy. We have no indication whether in the extreme south, opposite the Betic cordillera, a small amount of restituted energy has displayed its effects. If this latter deformation did indeed occur, it was weak.

The depression that characterizes the Parisian segment is, besides, well indicated on the other side of the Bay of Biscay by the axial dipping displayed, between the Pyrenees and their Cantabric extension, by Mesozoic deposits that predominate for more than 100 kilometers of length. Farther away, the depressed transverse alignment runs between the Paleozoic blocks of the Sierra de la Demanda and the Sierra de la Virgen. Farther to the south, most of the basin of New Castile still belongs to the depressed region. To whatever cause one may attribute the opening of the Bay of Biscay and the origin of its great depths, it is inevitable that the basement folds and their major features should precede all that; it is necessary that France and Spain, over the present location of the bay, be united in a single block. It is interesting to point out that in the hypothesis of the collapse and the fixed position of the continental block, many features on both sides remain somewhat unrelated. On the other hand, considering continental drifting, it is easy to observe that when the Spanish continental slope is brought back to the alignment or to the vicinity of the French continental slope, the Spanish basement folds take on a direction somewhat to the south, a direction that is very close to that which prevails in the other basement folds of Europe and that is identical to that of the Armorican basement fold; whereas the Parisian transverse alignment takes on in Spain the same general direction that it follows in France.

The great axial culmination of the Spanish basement folds occurs opposite the Hebridian massif, regardless of any idea concerning the mobilism or fixism of the continent. The same is true, as we have seen, for the Armorican massif and for Cornwall; it is true also for all the Highlands of western and northern Great Britain, which in Scotland are particularly raised in mass nearest to the obstacle. It is clear that during the times in which this powerful culmination of basement folds was developed from Seville to the Highlands, the Hebridian massif must have been more extensive and more resistant than it is today. The Laurentian massif was there, or whatever was connecting the Hebridian massif to the Laurentian massif. Besides, it is (267) probable that the usual axial behavior in the British portion of the system has not been everywhere developed to perfection because the influence of the extensive western resistance, today disappeared but in the past capable of deviating a portion of the flow lines from west to slightly northwest, is not inconceivable.

Such is, in its major aspects, the great Alpine deformation that has kneaded, folded, and dislocated the indurated basements of old Europe. Naturally, one does not have to go back to the times during which the Harz or each chainlet of the Great Basin was interpreted as an anticline. My distinction of basement folds and of voussoirs of basement folds takes care of that. The exposed horsts can be much more narrow than the basement folds in which they are generated. In contrast, let us recall the time in which each swell of the Rocky Mountains was considered a horst.

More or less framed within wide basement synclines or within depressed spaces between raised voussoirs, numerous cover folds in Europe have been active during Alpine times, and many display the effects of a small Andean phase. It is not necessary to insist on these deformations of which I have spoken elsewhere and which in most cases are minute repercussions of basement folding.

XXV. New Generalities on Basement Folding

This is the appropriate place to make some remarks of general significance.

In all the continents, including their associated blocks, basement folds predominate by their tonnage, and even more by the energy they consume, over the new chains, whether the latter arose from geosynclines or simply from monoclinal slopes facing the open sea.

Basement folding is the folding of the continental mass itself, from its plastic depths to its higher zones, where a lower plasticity predominates in varying degrees. A more regular deformation occurs at depth, which also is less unequally distributed than deformation in the upper parts. In the flowing that I have designated as *plastic flux,* the horizontal direction predominates. The heterogeneous upper parts, including subsurface and surface, lag behind and adapt themselves the best they can by borrowing energy from the deep flux; hence basement folds, their reactivations, their warpings, and, by means of a new derivation in place, their fractures. Thus, the energy involved deforms the entire mass of the continent, and no longer only the fresh and very plastic filling of the geosynclines. The deformation of these new chains results, on the whole, from a very modest borrowing of the energy that operates within the mass of the continents.

Until now I have presented no hypothesis concerning the origin of the flux: I have limited myself to detecting its distribution in space, its variations in time, its effects on basement folding, and its effects on the folding of new chains. (268)

Bailey Willis was the first to use on a large scale the concept of flux in his study of 1893 on the mechanics of Appalachian foldings.[21] Flux, a real undercurrent, is considered there as the result of a requirement of isostatic equilibrium; it is assumed to flow horizontally from the subsiding substratum of an area overloaded with sediments to the rising substratum of a zone unloaded by the progress of erosion. Later on, Bailey Willis presented a theory of Asia[22] that relies on similar assumptions. Rather recently, other authors have used the concept of undercurrent with a variety of theoretical systems. Until now, I have used the concept of flux only as a hypothesis that by means of a continuous operation is able to account for observable conformations obviously related to each other. Hence, I have preserved complete freedom with respect to the manner in which the flux itself may be explained.

In most cases, the existence of basement folds eliminates original and pure vertical movements and makes them very problematic in other instances. Therefore, the consequences of this change of type of deformation are certainly going to make themselves felt in all instances in which epeirogenic movements and other original vertical movements have been proposed; for instance, in stratigraphy and in geomorphology. It should be recognized that in many cases a simple transposition of language is sufficient; namely, what was said about pure vertical movements of continental blocks should from now on be understood as applying to the vertical component of basement folding and its secondary effects.

It is this component, together with other factors, that regulates the conformation and the arrangement of the substratum of transgressions and regressions. The same deformation, whenever of rising nature, starts the geomorphological cycles with their topographic rejuvenations and related conditions. This deformation also, whenever of sinking nature, favors the accumulation and all the phenomena of continental or marine subsidence. These alternations, while expressing in a final analysis the behavior of a horizontal effort, generate those of the topographic conditions.

In a famous contribution,[23] Bailey Willis clearly established the concept of a very young age for most of the reliefs of Asia. This is also true, as we well know, of a very large portion of the other emerged lands. What I have just said concerning the vertical component of foldings appears to me to establish the true relationship between the topographies consisting of several cycles and the horizontal efforts that have continuously acted—by compression or by traction—within the mass of the continents and their associated frames.

Furthermore, I have continually used the term "*tangential* effort" as synonymous with horizontal effort or with predominant horizontal deformation, without any theoretical implication whatsoever. On the other hand, I have included in the designation "radial" all the theoretical considerations related to it during the history (269) of our science: namely, original, pure vertical movement and the concept of the large-scale sinking of continents.

The existence of a certain residue of original vertical movements remains highly problematical because if one extends to the extreme the interpretation of facts, there seems to be no tectonic deformation, not even one exactly vertical, that cannot be considered as the expression or the direct or indirect consequence—near or remote, with a relatively short delay or a very long deadline—of deformations in volume in which horizontal deformations prevail or have prevailed.

First, it should be pointed out that in a medium with any degree of plasticity, no pure displacement, as a whole and along a single direction in space, is possible. There is necessarily a deformation in volume, and strictly speaking there are no more pure horizontal movements than there are pure vertical displacements.

For objects of small extent, for little voussoirs of the immediate subsurface whose deformation demonstrates the amount of rigidity that can subsist locally, one talks preferably of movements as a whole in one single direction that often is vertical or almost vertical. But unless we restrict ourselves to the smallest scales, this manner of speaking remains ambiguous and only valid in a first approximation.

This manner of speaking becomes increasingly imprecise and generates more potential or obvious mistakes whenever objects become wider and thicker. A tectonic object of appreciable extent cannot undergo displacement, in any direction, without being deformed. This is perfectly clear for the horizontal aspects of deformations since a finite push cannot be transmitted in a plastic medium without weakening with increasing distance and without eventually becoming exhausted in the internal frictions of the mass. This is equally true, although in a slightly different sense, for the ordinary case of unequally plastic media that fill space, that influence each other and along whose limits discontinuous exchanges of energy take place, and that favor in many cases a conduction, a reconduction, or a more distant propagation of partial efforts. It is even clearer, if anything could be clearer, for the vertical aspect of the deformations. No tectonic medium, beyond a certain thickness, can support its own weight. And regardless of how confined the matter is, it is not conceivable, in general and strictly speaking, that the ascending or descending trajectory of a given point be everywhere and always exactly vertical: a great number of factors are present to impel large or small lateral deviations.

The concept of pure displacement, so frequently implied in tectonics with respect to twenty kinds of movements and to a multitude of concrete objects, is hence never manipulated with enough care. This concept should be kept as an analytical artifice only. Really, to conceive a pure displacement (270) is in general nothing else but to visualize, at one given point and for an instant of time, a tangent to an infinitesimal constituent of a curvilinear trajectory. It is only in the very particular case in which the trajectory is rectilinear that the artifice becomes reality at a given place and for a certain time. According to current usage in tectonics, the arbitrary position consists not only of replacing a curve by a particular tangent but also of speaking as if this tangent, horizontal or vertical, were the trajectory itself. As anyone can see, this arbitrary position is more than implied. It hides itself implicitly at the very depth of the thought, and the error it conveys remains too often undetected, while the real nature of the displacements and of the deformations remain similarly undetected.

Therefore, it is sufficient to restitute to space its plenitude and to think of the nature of plastic deformations—in order to make the concept of pure displacement vanish and to deprive it of all reality except in the very particular cases in which such a concept is merely acceptable—as a simplification of language and as a first approximation.

Let us think of the hypothesis of eustatic movements and of the rather simple and equalizing geometry that is inseparable from it. Certainly, nobody can refuse to admit some temporary or permanent changes in the volume of the liquid mass of the oceans. The formation and melting of the great *inlandsis* have been considered, with good reasons, as related to temporary and limited changes in the oceanic volume; I shall say nothing about other factors that have been considered except to state that some of them are plausible. With respect to variations in the capacity of the vessel that contains the liquid mass, everything seems to indicate that they are very slow to occur. At best, without forgetting the general deformations of the geoid, one can see that there will be no eustatic movements but only vertical hydraulic effects that could, in a first approximation and in a rather risky manner, be assumed to be eustatic, that is, equally distributed along all shorelines. These hydraulic effects certainly will be combined, by addition or subtraction, with the vertical effects of the proper tectonic deformation, effects that are very complicated and very unequally distributed. The elevations of shorelines could not furnish—through their readily available data only—the means to separate the hydraulic effects from the tectonic ones and consequently to demonstrate the eustatic character of the former. Whether it is possible or not to detect by means of a further elaboration of the same data hydraulic effects operating as a vertical component in the general vertical effect—the only directly observable reality—is a question I shall not discuss now. To these difficulties of the theory of eustatic movements can be added, for the lowermost terraces, the fact that the original differences of elevation of the wave-cut surfaces, or of the marine beaches, are of the same order of magnitude as the vertical effects of which one tries to measure with precision the amount and the nature. (271)

Considering foldings of any kind, I have shown what pertains to the vertical aspects in their behavior.

If one considers the movements of great domes, huge shields, or of extensive platforms, one has even stronger reasons, on such a large scale, for speaking about displacements as nothing but aspects predominating or not within a deformation from which they are inseparable. What kind of displacements in this deformation of great style are dominant or predominant or have been so because of variable time and space circumstances? Contrary to first appearances, they are the horizontal displacements. It is time to discuss them.

Because basement folding has reworked, as we have seen, all the zones of dead folds that border or surround these much older massifs, there is no reason to assume that it should have avoided the latter, themselves made up of still older dead folds. If such a situation were assumed, it would mean that the oldest massifs have zero plasticity, and all contact with reality would be lost. Unquestionably, as stressed many times previously, the plasticity of these ancient nuclei is, on the average, lower than that of less older dead folds; but all this is a matter of degree and not of difference in nature. Consequently, we cannot avoid attributing to basement folding the essential features of the behavior and of the conformation of the broad structures with very large radius of curvature, namely, the broad domes, the extensive shields, and the great platforms. These objects are or have been basement brachyanticlines or complexes of basement folds. Their lesser plasticity, which does not prevent them from undergoing basement folding, nevertheless gives them in most cases a particular manner of reacting. The predominance of the large radii of curvature originates from this situation. These same objects behave as buttresses for other basement folds, generally better ordered and consisting of less ancient materials; they behave again as buttresses for new chains, under rather different conditions. Nevertheless, in all cases, they are *buttresses that fold*. At times of well-characterized compression, these buttresses have more than once been warped while rising, in the same manner as the other foldings, either new or of basement type that belonged to the same segment. Furthermore, their resistance with respect to the former and to the latter became thus increased in the upper parts, a circumstance that in turn has influenced the rising of the basement folds and of the new folds facing them, by reinforcing their axial behavior. This reciprocal causality in the behaviors of objects that resist while folding, with degrees of resistance as well as degrees of propensity for being folded, is one of the major features of the deformation of continents. All this should not be interpreted as the expression of an essential difference in the nature of the movements—but quite the opposite, in fact. The numerous reactivations, so many times repeated in the same style that I have described, were more accentuated because of such conditions.

For extensive areas with very large radius of curvature, as well as for regions with well-ordered basement folds or for new chains—in fact, for all kinds of lands (272) built by somewhat large-scale structures—we have three orders of facts whose synthesis is one of the major problems for the future: observable tectonics, gravity anomalies, and isostatic behavior. Many data are still lacking, particularly with respect to the last two items. However, until just recently, useful attempts have been made to unravel the complicated interactions that characterize this problem in both its broad lines and its regional aspects.

The interest displayed in comparing the gravity anomalies with the observable geological data depends to a great extent on the kind of reduction applied to the raw gravimetric data. In this respect, the difference between gravity reduced according to Bouguer and the normal theoretic gravity at sea level presents, compared with other modes of expression, the advantages of being freed from the attraction of the masses rising above this level and of being referable to the influence of the deeper masses. In numerous cases and by taking certain precautions, we may find profitable the geological discussion of anomalies calculated by other methods.

Positive or negative gravity anomalies that were corrected or not are interpreted as related to the underground distribution of densities. According to one hypothesis that is not always accepted but still has appreciable merits, the anomalies are related, to an extent often considered as fundamental, to the variations in thickness of a light material, the *sial*, beneath which the much denser *sima* is supposed to occur. Actually, several authors tend to consider the anomalies as representable without using this hypothesis, but I shall not indulge in this discussion. If one believes this hypothesis to be valid, it is not necessary to think that the sial and the sima display everywhere sharp boundaries—but this is a mere detail.

Isostatic behaviors are related to tectonic deformations proper among which folding appears to play a prominent role, or to phenomena of erosion or sedimentation, of glaciation or deglaciation, or to the combined and variously weighted effects of all these factors and their repercussions. In more or less complicated interactions, these different factors may appear as cause, as condition, or as effect of isostatic behaviors related to variations of gravity and to more or less horizontal subcurrents of which many types have been visualized. Whenever these subcurrents are assumed, as in the ordinary case typical of many recent systems, to flow from an overloaded and sinking substratum to a normal substratum or from a substratum of either one of the two types to one that is unloaded and rising, one is really facing small-scale variations of the concept put forth by Bailey Willis in 1893,[24] regardless of the causes presented to explain the overloading or the unloading. In the hypothesis of a superposed sial and sima, all the preceding factors can furthermore (273) act by means of deformations proper to the sial, with changes in the thickness of that matter, and by means of movements that concentrate in the sial, in the sima, and in their varieties as a consequence of the different properties of these media. It is readily apparent that this sketch because of lack of time was purposely limited to a few general aspects and appreciably stripped of fine nuances that are peculiar to observed behaviors. Therefore, it is still very far from encompassing the entire complexity of actual events.

The new adaptations satisfy to a variable degree and with a variable time lag the requirements of isostasy. Very often, a new isostatic behavior prevents an older one from reaching complete development or covers and even destroys the effects of the latter, replacing them with its own.

One should naturally consider as very much derived, in other words, as *secondary* or accessory, the gravitational and isostatic behaviors regulated by external agents; for instance, those generated by the overloading from sediments or large glaciers and those initiated by the unloading due to erosion and ice melting. Indeed, without something being deformed or having been deformed, one cannot conceive of either erosion or accumulation. It is also difficult to visualize the greatest vicissitudes of a very widespread glacial regime without movements belonging to a deformation. These accessory behaviors therefore imply perturbations brought to previously stable or unstable conditions. It is clear that they belong to a process of a much greater scale, which is based on deformation and gravitational and isostatic changes directed by internal actions; that they relate to it as subordinated consequences—being responsible for new detailed effects; and that they certainly do not dominate this much greater process to which we are necessarily led.

In this difficult inquiry, nothing will be gained by simply assuming a varied and repeated functioning of these secondary mechanisms throughout the entire past; indeed, the derived effects will never be able to account for the major process.

The hypothesis attributing to the first crust of the sial, developed in pre-geological times, the main variations in the thickness of that matter, does not satisfy the mind any better. It seems natural to think that after so many episodes of reworking, nothing remains of the original shape of that crust.

Therefore, should one be limited to the diagnosis of secondary behaviors, in part directed by external factors, or to the elucidation of the isostatic and gravitational behaviors of more direct origin, displayed by the new chains and their marginal depressions? Should one attribute to this group of factors that are, besides, extremely complex, all the causes of the redistribution of matter that lead to the major gravity anomalies and to the major variations in the thickness of the sial? No, much more than that is available: one has the process of basement folding, and one can furthermore take into account its complementary aspect, namely, the distensional traction of the sial (274), which is after all the negative aspect and the required corollary of basement folding.

It has been known for a long time that in geosynclinal folds, and quite often in cover folds, the thinned parts alternate with those showing an excess of material. Limiting ourselves to a few examples, let us briefly recall that the overturned limbs of recumbent folds usually display the former behavior and their frontal archbends the latter. The alternation of thinned and swollen parts is the rule in the most deformable beds of cover folds, although the alternation is of a different style than that displayed by recumbent folds. *Now, basement folding, positive or negative, can be considered as a cover folding developed on an immense scale.* It is the behavior of a covering, consisting mainly of the sial, that becomes complicated within and over the sima—by taking advantage of the variations in plasticity it displays in the vertical direction, in the horizontal direction, and in all directions, by stretching in places and during a certain interval of time, by renewing these behaviors and letting them alternate in a varied manner, and by gradually yielding, because of its great thickness, to the solicitations of gravity and of isostasy, to a degree that cannot be visualized for cover folding proper.

If this interpretation is correct; if basement folding should be conceived of to such a degree of generality; if it is supported by all these factors, then one has additional reasons for admitting that it characterizes not only zones consisting of ordered basement folds but also areas formed by widespread structures with a very large radius of curvature: broad domes, extensive shields, and old platforms. Basement folding forms and deforms all these objects; it is the main aspect of a fundamental process that overtops and commands the basement folding itself and, consequently, the deformation of new chains and cover folds. In fact, this large-scale process directs—by means of foldings of all kinds, or through other paths, or through less indirect ways according to circumstances, and with or without the help of external actions—the secondary behaviors that are similarly based on deformation and on gravitational and isostatic changes. The behaviors generated by erosion, sedimentation, glaciation, deglaciation, as well as many others directed by purely internal causes, are therefore only secondary consequences grafted upon the main process, which they modify in places and for a certain length of time.

Today, we can say only that *primitive* conformations of the base of the sial did exist. But these are conformations that may be qualified as *essential* or original and that in spite of their ceaseless variability can be related to the general deformation of which basement folding is the most visible expression. The network of conjectures by means of which one tries today to build a deep-seated tectonics, an *infratectonics* that would account for the three orders of facts, (275) is already very complicated. After what has just been said, it would be necessary to add new working hypotheses relating on the one hand to the influence exerted by the basement folding on the deep-seated conformations, on the gravity anomalies, on isostasy, and on the behaviors that express the numerous relationships among these factors. On the other hand, these new hypotheses should also relate to the counterinfluences exerted upon basement folding by the aforementioned features.

One can visualize beneath the large basement anticlines, as well as beneath the new chains, the formation of *lenses* of deep-seated sial. These lenses are more or less imperfect negatives of the main visible intumescences through which isostasy is more or less satisfied and with respect to which these intumescences show, as they still do for the new chains, an overcompensation in the case of rapid growth of the folded wrinkles, an undercompensation in the opposite situation, or an approached compensation. However, tectonic complexes of which certain elements are overcompensated and others undercompensated will nevertheless show, with respect to the entire area, an *interdependent* compensation due to the intervention of elastic or semi-rigid effects. Besides, all things being equal, this interdependent regime will have a greater chance to become established in a complex with basement folds, rather than in a complex with new chains, because of the greater degree of induration of the upper parts of the sial in the former case. The total excesses or deficiencies of mass will imply, to a certain extent, a work performed against gravity or against the buoyancy force and more or less unstable conditions, except in the case of interdependent compensation. A particular case of instability will develop with the same reservation when the anticlinal basement folding operates in a region of positive anomaly; for instance, a sial whose thinning is not compensated. More or less horizontal migrations of deep-seated material, sima or sial, will occur along directions regulated by the excessive, normal, or deficient overburden that acts on the particular deep-seated area. Thus will be outlined the beginning of secondary redistribution of the materials and of the densities, with new gravitational, isostatic, and proper tectonic effects. Clearly of secondary character but still related in the final analysis to the large-scale general process or to its regional aspects will be the behaviors generated by the external unloadings or overloadings, with the changes they will bring to the subcurrents and to the migrations; with the imports and exports of sial or of sima that will be inseparable from them; with the new isostatic rising and sinking that will result from them; with the redistributions of gravity anomalies related to them; and with all the subsequent behaviors of similar nature but more indirectly derived that will occur.

Under such conditions, one would not expect that the original effects of basement folding or of the folding of new chains (276) would be preserved everywhere without alteration. Quite often, one should take into account very important secondary reworkings. The infratectonic conformations will not necessarily be the negatives of the visible intumescences. The interpretation of the maps of isoanomalies by means of the visible tectonics will require numerous precautions and a great visual ability to distinguish the essential from the accessory. Therefore, the inverse interpretation will be even more delicate and very risky. The present distribution of the anomalies depends upon the infratectonics and therefore on the total tectonics, which implies the survival of many features related through a thousand interactions to previous states of the system. Besides, all things being equal, one should expect that the part of the original deep-seated conformations would be the less discernible, the smaller the tonnage of the basement folding and the more important the secondary reworkings. The inverse variations of the same factors, namely, basement folding of great tonnage and absolutely or relatively weak secondary modifications, are favorable to a more or less excellent preservation of the major deep-seated systems. They will be expressed by a better fit between the visible tectonics and what the maps of isoanomalies tell us about infratectonics.

These criteria being established, we can see that large portions of old Europe with weak basement folding and important secondary deformations seem to recall the first case; whereas the extensive regions with powerful basement folds, such as the western chains of both Americas or the segment of central Asia, recall the second case. In the present state of measurements, a control is possible in certain regions only, but it is very instructive.

Among the numerous maps of isoanomalies of the United States prepared by different methods and recently published by W. Bowie, we shall discuss the one that alone can display, in regard to the visible tectonics, relationships that can be understood: it is the map that corresponds to the reduction of Bouguer.[25] Most of the negative anomaly is located under the major part of the western ranges, that is, under the Andean-Alpine and particularly the Laramide basement folds. An extensive lens of deep-seated sial, which shows with respect to the great visible intumescences several features of acquired incompatibility, is therefore admissible. I shall discuss later on this incompatibility, which results from a fundamental factor and is certainly not of secondary importance. The original and approached compatibility nevertheless remains easily discernible. The association of the chains and of the lens is, in a sense, a basement fold of high order, reworked again and again in several phases. The Colorado Plateau, in spite of its being an old shield and despite (277) its large radii of curvature, is no exception. The Alpine deformations that led to the rising of the Colorado Plateau as a single mass during the Neogene or at the end of that period represent essentially basement folding, strongly accentuated by a horizontal effort and emphasized by longitudinal fractures particularly well marked in the region that is most exposed to stresses—the western promontory. Moreover, several generations of fractures occur.

In order to define more precisely within the entire intumescence of the chains, on a scale immediately below—which is that of particular basement folds—the degrees of compatibility and incompatibility between each of these objects and the data of the infratectonics, we have to wait for an increase in the density of the network of stations that will allow us to draw more detailed maps of isoanomalies. The isoanomalies available at present can express only the most generalized conformation of the lens, and nothing more, because of the small number of stations and the great distances between them.

The system of old Caledonian and Hercynian folds that build today the Taconic chain, the Piedmont, and the Appalachians has undergone a large-scale Alpine basement folding that has had infratectonic effects, expressed by a negative lens; tectonic effects, visible in volume in the general intumescence of the relief, and at the surface in the repeated warpings of the classical peneplains of that region; morphological effects by rejuvenation of the topography; stratigraphic effects, complementary to the erosion, through sedimentation in peripheral positions; and more minute effects belonging to any of the preceding types and whose complicated interactions are just beginning to be unraveled. These minute effects may be said to depend, as accessory behaviors, upon the major processes just described, and eventually, for their essential part, upon basement folding. In the latter, an Andean phase has taken place, probably synergic with the great Andean movements of the then new chains of the western United States. This situation is demonstrated by the unconformity of the Early Cretaceous beds of the Potomac over deformed Triassic, which has been observed in a few parts of the Atlantic slope of the basement folds. The other stratigraphic gaps displayed by the Cretaceous and Cenozoic deposits of the coastal plains will in most cases be conveniently related to the renewals and progresses of the basement folding or to its more complex repercussions.

Most of the negative lens, which is of appreciable size although much smaller than that of the chains of the Far West, agrees approximately by its location with most of the visible relief, that is, the Appalachians. The incompatibility that exists on the largest scale has the same direction, is of the same kind, and certainly has the same origin as that of the Laramide and Andean chains. The western continuation of the lens expresses the extension of the Appalachians into Arkansas and Oklahoma; it is indicated by a few stations with relatively high negative values. (278)

The two main regions of ordered basement folding, namely, the great western ranges and the eastern United States, thus verify in a suitable manner and in spite of alterations that do not pertain only to details the reality of the essential and original behaviors that I associate with basement folding.

The maps of isoanomalies of the United States, based on reductions other than those of Bouguer, show very few clear relationships with visible tectonics.

The huge intumescences, with a more or less meridian direction, of East Africa and of both sides of the Red Sea may be considered essentially as a system of basement folds with a predominantly anticlinal character and related to a compression acting in the direction of the parallels. The large submeridian fractures, as impressive as they might be, are reduced to simple details in a deformation of such tonnage, and many of them can be classified as longitudinal fractures related to the basement folding. This does not mean that the latter process was the only one active and that one should neglect the hypothesis of horizontal tractions that operated along the direction of the parallels. Assuming the succession or the alternation of folding and stretching deformations, one can easily conceive that the latter may have generated other meridian fractures, of distensional origin, and that the fractures of one or the other origin, which were so to speak taken advantage of by the continuous traction, may often have been enlarged into these great grabens, which represent one of the most striking features of that part of the Earth.

This complex working hypothesis, more flexible and richer than the exclusive theories, has a much better chance of accounting for the requirements of the visible tectonics. In the domain of infratectonics, this hypothesis allows us to eliminate several difficulties. If there is basement folding, one can expect to find, if no major reworkings have occurred, noticeable negative anomalies under the great intumescences: this is indeed the situation revealed by gravity measurements over large areas of East Africa. Whether under such conditions the grabens are close to being compensated or not, is a problem that belongs to the scale immediately below characterized by behaviors of partly secondary nature as well as by factors from which, at the present time, there is no reason to remove the distensional deformations and their consequences.

Extra-Alpine Europe, by the small-scale complication of its tectonics, by the weakness of the visible basement folding, by the presumed weakness of the subcrustal features originally related to this folding, by the importance of secondary reworkings of all kinds and of more general perturbations, is probably, of the entire Earth, one of the regions most prone to hide the essential under a profusion of details. I do not say that this is the case for all parts of old Europe. But these circumstances, many times repeated without losing (279) the rather widespread character just mentioned, will in many places prevent us from reaching a conclusion—without being somewhat arbitrary—about the original shape of infratectonic objects that are today offset, altered, reworked, or completely destroyed. Therefore, it will not always be easy to reconstruct the essential, as may be done in countries built on a much broader scale. What is shown in these regions by the maps of isoanomalies is often but such a destruction, more or less advanced, even completed or exceeded, under influences, partly secondary in nature, that managed in time to dominate or even completely to take over the situation. This is the reason why during these short minutes, I cannot analyze any further this problem which is of interest only when discussed in detail on the lower scales.

However, I shall mention the recently proposed hypothesis of the suction of deep-seated material through horizontal subcurrents toward the unloaded and rising basement of the recently deglaciated central part of Fennoscandia. This suction has certainly replaced inverse repulsions generated by the last glaciation, and the two types of perturbations must have alternated during earlier Pleistocene times because of periodic glaciations and deglaciations. This allows us to visualize the importance of a portion of the secondary reworkings that have affected the deep-seated conformations related to broad zones of basement folding situated in peripheral positions with respect to the Baltic shield. I can also cite the small intracontinental furrows, sometimes called geosynclines, in which are recorded different isostatic and gravitational processes influenced by sedimentation and which, in comparison with geosynclines proper, display several differences in behavior. Besides, these different processes are far from exhausting the inventory of the secondary perturbations recognized at present.

I shall not discuss here the influence that the interdependent compensation, for which more than one part of old Europe shows a propensity, can exert on the diagnosis of the differences between the secondary conformations and what remains of the less altered essential conformations.

Central Asia, with its huge basement folds, has a greater chance to show a predominance of more original deformations. Therefore, the measurements of gravity pertaining to several parts of the Western Tien Shan, to the adjacent plains, and to the Pamir give strong negative anomalies, which imply most probably powerful lenses of sial protruding from the deep-seated part of the mountains into that of the plains. In this respect, most of the segment of central Asia remains temporarily unknown, but its gravimetric exploration will certainly disclose important facts.

It is superfluous to recall data pertaining to the Himalayas and to the Indo-Gangetic foredeep, which are classical regions with respect to the history of gravimetry. But the fact should be stressed that the mountainous intumescence (280), within the limits of the territory thus investigated, results essentially from a basement fold, namely, the Himalayan zone, and contains a relatively small amount of new tonnage. Consequently, the essential isostatic and gravitational behaviors in that region are those appropriate to basement folding, and in general they should not differ much from those generated by the formation of new chains. The few other regions with large basement folds that we have just examined give the same result; if certain areas of extra-Alpine Europe may suggest the contrary, it is perhaps because they have been considered by themselves and also because the accessory has been taken for the essential, and the particular for the general.

The horizontal migrations of deep-seated material are expressed by vertical effects that consist, except for inversion due to other, more powerful factors, of a lowering in cases of the export of material, and a rising in cases of the import. It is conceivable that these effects may be reflected in somewhat important alterations of the shape of the folded objects, reaching their visible parts, and in new fractures that develop in the upper parts. One is led to think that more than one regional volume of basement folds or of new chains has undergone secondary deformations of this kind and, therefore, more or less well marked obliterations of the original arrangements. Conformations due to original axial behaviors, in particular, may become altered in such a manner. One should certainly take into account this kind of secondary perturbations, and each particular case should be examined by itself. But with this type of accidents, one should not try to explain most of the other conformations as a whole, particularly on the higher scales; it would be like producing obscurity where clarity reigns and would properly lead to chaos. The reason for this is obvious. In a great majority of cases, as just shown by my investigation extended to the most important folded areas of Eurasia and of other continents, the great visible axial behaviors satisfy conditions imposed by the interaction of new or basement folding with obstacles opposite which a given vertical effect takes place. This fact is of capital importance; it shows that axial behaviors are not distributed at random but follow rules that relate to the immediate conditions, to the most intrinsic behavior of the folding, new or of basement type. Therefore, on the whole, secondary effects of any kind are very far from being predominant. In a striking manner—by a rule of statistical frequency, a rule of dominance—this fact summarizes the result of a continuous fight during geological time, a fight probably still occurring today, between the great folding processes of the sial and the secondary deformations. Therefore, it would be difficult to give to this predominance of the essential effects an expression more condensed, more synthetical, and more implicitly rich in all the heritage of past processes. Considering the duration, one can see that this dominance results on the one hand from the fact that the secondary behaviors (281) rarely succeeded in imposing themselves—at most they did so for a certain time and in particular regions only. On the other hand, it results from the insistence, the rapidity, and the success with which folding, continued or restarted, usually succeeds in repairing these temporary deteriorations, in creating new essential conformations, in liquidating secondary obliterations, in dominating these accessory behaviors, and in letting its own laws prevail.

It is this rule of frequency, this dominance of the essential effects, renewed or not, that gives to the consideration of the gravity anomalies, in the light of the visible tectonics, a great interest. The dangers often presented by the inverse process, by means of which one attempts to interpret the visible tectonics through the anomalies and through a conjectural infratectonics in order to explain the best known by the least known and even to dispute sometimes the most immediate consequences of visible geological facts, these dangers result, on the one hand, from the multiplicity of possible explanatory combinations and, on the other hand, from the omission of the data of the concrete tectonics.

The rule of dominance of the essential effects is true for all the large continental masses: it may suffer a few exceptions in particular areas in which folding has not regained the necessary energy for an efficient renewal, or in which secondary effects can predominate until such a renewal, or in which they can be visible, meanwhile, in the infratectonics and as high as the visible tectonics.

More radical exceptions can occur, for instance, in the case in which disjunctions, incipient or completed, render the renewal of folding precarious or impossible. In a few moments, I shall discuss this question, which nevertheless had to be briefly mentioned here.

The same statistical law stresses axial behaviors, in other words, what is most important in the vertical aspects of the general deformation. In this manner it establishes—by calling upon a very simple relation with the obstacles it faces—the causal subordination of most of the vertical aspects to a plastic flow that is essentially horizontal. I have verified this dependency in a hundred concrete examples, and it is superfluous to recall this fact.

The analysis just completed allows us to organize a great number of materials for a more encompassing synthesis.

First, the general tectonic deformation, which directs everything else, is mainly and visibly expressed by a large-scale behavior in which the horizontal aspects predominate. It is basement folding with its positive aspect expressed through compression, and its negative aspect expressed through traction. The general tectonic deformation is furthermore expressed by the folding of the new chains whose character is nevertheless already somewhat secondary since the deformation involves a relatively small reworked tonnage of plastic sediments and not (282) the major portion of the original sial. These different folding or stretching behaviors imply extremely varied vertical effects because of time and location factors. These behaviors therefore do not permit any certain diagnosis of displacements, altogether pure and original, of movements as a whole, whether vertical or not. Hence, the dominance of the horizontal aspects is even better established.

Consequently, it is clear that the concept of basement folding renews, from base to top, large portions of the structure generated by the ancient tectonics: continents that bend or become deformed throughout their entire mass; geosynclinal chains that are reduced to the rank of details; vertical movements pure and original that can be dispensed with. These movements being no longer distinguished except by artifices of analysis or precisely separated from the vertical effects of the tangential stresses—effects that are essential or accessory, direct or indirect, tectonic or infratectonic—it becomes impossible to state anything positive about them. As to the essential vertical effects, these are immediately obvious; with respect to the vertical effects of accessory behaviors, the evidence appears only at the end of a more delicate discussion. The concept of vertical tectonic movements that would be altogether pure and original implies a postulate not demonstrable in theory, and essentially superfluous in practice.

Underlying all this is the helplessness of linear schemes to account for any deformation whatsoever. The mind takes hold of these schemes, finds them simple and flexible, and considers them, not without reason, as better adapted to the texture of language than to the inexhaustible richness of the concrete mobility within concrete space. These economical schemes become in time a kind of common currency whose value is no longer questioned. But, men who are sensitive to the requirements of the concrete, men who are endowed with a strong visual tendency, cannot be satisfied with these schemes. This situation has led to many renewals of the structure of our science. Among them, one of the most successful has been in the past the substitution for the idea of axes of uplifting the ideas of folding and of horizontal compression. The concepts of epeirogenic movement and of radial dislocation find themselves today in front of the essentially horizontal character inherent in the general tectonic deformation, in the same critical situation and devoid of all indisputable foundation. There are no vertical tectonic movements; there are only vertical effects that can be related to a deformation in volume.

Of course, I do not intend to ignore the influence of gravity and of the vertically oriented forces that it generates. This entire discussion indicates such an attitude. But a component which is an abstraction or a product of reason is one thing, the tangible and concrete effect that I wanted to stress is another.

It is also appropriate to add that by these considerations concerning the pure and original vertical tectonic movements, I do not intend (283) anything about deformations of another kind, such as the general deformations of the geoid. Similarly, I do not plan to extend them to the vertical behaviors that geophysics and geology often accept as a hypothesis in deep-seated and more or less viscous fluids, such as vertical subcurrents, the sinking or rising of immerged bodies, the vertical distribution of materials in order of density, and so on. It seems to me correct to apply these considerations, for the time being, only to the deformation of the plastic media that usually are directly involved in tectonics.

Second, the predominance of the horizontal aspects in the greatest deformations that I have recognized shows that the same should be true of the even more general tectonic deformation from which all this depends. Therefore, the synthesis of the observable, on the scale immediately above that of basement folding, is conceivable only through a hypothesis based on essentially horizontal behaviors. Furthermore, these behaviors, in order to explain the general distribution of basement folding, must encompass the entire mass of the continents and of their adjacent parts.

To arrive at this conclusion, it was not necessary to use such ideas as have been put forward concerning the fixism of continents or the ability of continents to undergo considerable horizontal displacement. Therefore, the preceding remarks are independent of these theoretical viewpoints.

None of them, strictly speaking, implies the necessity of a choice between these different points of view. But right now, we can clearly see them leading to a view that agrees well with the second hypothesis. I shall now examine for a few minutes this question, which has so many fundamental consequences.

XXVI. Continuum and Discontinuum in Tectonics

The need to arrange the inexhaustible aspects of nature, which operates in the three dimensions of space and time, into the linear order of the talk left me only with the choice between either a chronological distribution of the narrative or a subdivision by regions. I have subordinated the second arrangement to the first as best as I could. In this manner, I have remained as close as possible to what nature can offer in terms of continuous visual appearances of interdependent aspects cast in a single episode and also as close as possible to a view of things that I can frankly say is fruitful and convenient. This view consists of accepting the discontinuous wherever it occurs and integrating it into the continuous on higher scales.

The discontinuous behaviors are obvious wherever their trace has been observed. The position of the continuous, an act of synthesis that overrides analysis in order to recreate it more rich and harmonious, is the unity of complexity: I am not supposed to discuss its titles or to investigate its essence. It is sufficient to say that in tectonics nothing can be invented, on a large scale, without a visual imagery (284) of the continuous deformation; that this imagery can lead to serious miscalculations or to the greatest aberrations when it is not checked by a criticism that is inflexibly rational but nevertheless endowed with insight; that too much criticism kills the right invention, and too little lets error loose; that this imagery fulfills its role when it has been fruitful in verified discoveries; that these discoveries are after all the only perceptible thing; and that it is superfluous, when speaking of science, to ask what is in reality the concept of a continuum. Cataclasis and recrystallization demonstrate to what extent the appearance of continuous deformation, this extremely useful instrument of synthesis on a large scale, can be expressed by a summation of an immense number of small discontinuous movements involving minute areas or points.

Subordinating the regional order to that of time, I was able to avoid frequent retrospective glances, and I have accepted the obligation of ceaseless displacements. These increasingly extensive excursions of the mind to the east or the west used the segment of central Asia as a hinge. The previous excursion ended along the western shorelines; the next one shall take us back to the Far East.

East Asia is the association of tectonic objects in which the huge deformation of the Circumpacific chains, generally parallel to the shores, is complicated by lateral repercussions, by marginal deformations that the powerful frontal compression between India and Angara Land—concentrated in the segment of central Asia—generated farther to the east.

The association of these two major factors dominates the entire behavior of East Asia beginning at a certain point in the Alpine cycle. Earlier, the Circumpacific deformation, complicated in places by all the features brought about by the presence of ancient nuclei, was the only important process. But the directions, which are almost perpendicular to the coast and which occur over large areas, mainly within the Amurian lands and in a portion of the Sikhota-Alin, allow us to visualize the complexity of more particular conditions that may have prevailed in certain regions.

Present-day East Asia consists of a single continental mass, warped, shoved, and dislocated by Alpine basement folds, with numerous cover foldings variably framed; it consists of about ten marginal seas; of a series of festoons that in some places are attached to the continent and display within their structure of new chains, whenever the axial culmination is sufficient, important basement folds of dead ante-Alpine material. These last two circumstances indicate that the deepest part of the festoons belongs to the continental mass, from which it was separated in an incomplete manner, with the exception of small fragments.

The continental block of East Asia stretches from western Indonesia to the Bering Strait. It includes Indochina, China proper, Korea, Manchuria, the Amurian lands, the northern surroundings of (285) the seas of Okhotsk and Bering, and the submerged continental shelves and taluses that connect or border all these lands. This block is united with that of North America, in fact, the festoons of the Aleutian Islands and the powerful virgation of basement folds of the Alaskides, west of the 140th meridian, are as much Asiatic by their plan as American by their material. Among the marginal seas are included not only the five examples that extend from the Bering Sea to the China Sea but also the seas of Sulu, Celebes, and Banda, as well as the areas of rather deep sea that extend from the arc Andaman-Nicobar to the talus of the Malay Peninsula. The festoons follow each other from Assam to the Aleutians. The Circumpacific type of the deformations, opposite the great depths of the Indian Ocean, is as clearly expressed in the Burman and Malaysian arc as it is facing the Pacific, between the Philippines and the Aleutians. The fact that this arc has been met at both ends, by the old promontory of Assam and by Australia, is not obvious over great distances. The complex of arcs that encompasses the Izu Shichito Islands, the Bonin Islands with their folded Nummulitic and unconformable Neogene, the Mariana Islands, the western Caroline Islands with the crystalline schists of Yap, the Palau Islands, and a few other smaller objects in the southwest terminates the present Asiatic frame on the Pacific side. The Mariana Islands, the western Caroline Islands, and the Palau Islands are arranged in echelon pattern. The broad and deep space that extends westward of the complex of arcs, as far as the submerged foot of the cordilleras of southwestern Japan, of the Ryukyu Islands, of Taiwan, and of the Philippines, is in a certain sense the tenth marginal sea of Asia.

East Asia is thus defined.

The middle Caroline Islands with the crystalline schists of Uola and the eastern Caroline Islands belong to the delicate cortege of the Oceanides.

On both sides of the segment of central Asia, the deformations of the Angara continent and of its margins display analogies that result from a certain number of common conditions, similarities that attract suspicion and dissimilarities that present the most fundamental problems.

Most of the analogies are due to the fact that Angara Land and its margins of young chains were deformed, on both sides of the most compressed segment, into two protuberances convex to the south, which attempted to envelop India at both wings.

The first of these arrangements, represented by the southern half of the Turanian segment, includes the basement folds of the Turan and the new chains of Iran with incorporated basement folds.

The second arrangement encompasses the new arcs that extend from Assam to Taiwan, through Indonesia and the Philippines, and the old lands enclosed within that circumvallation. With the exception of the Burman arc, and without mentioning the very old and problematic massif that was supposed to exist in the southeast, Indochina is a bundle of basement folds that display a tendency to diverge toward the south and the southeast; cover foldings (286) have occurred in some places. The nappes of southern Yunnan, of Tonkin, and of northern Annam are basement folds that upon rupturing by large fractures—that have low dips and are repeated—have exceeded and perhaps lost all ordering in the mechanical behaviors belonging to the evolution of basement folds when it is carried to such a degree. These nappes seem to form the right wing of an arc whose left wing, which remains yet to be found, would stretch or would have stretched to Hainan and to the margin of southern China. In any case, what is known of their ground plan fits well into the family of curves displayed by the other great structural lines of Indochina and by the young arcs of the external circumvallation. This situation results from a certain solidarity of behavior that did not encompass the entire Alpine cycle everywhere, some dismemberments having affected several segments of the periphery. The basement folds of most of Indochina behaved, to a certain extent, with respect to the new arcs *downstream* from them—the Burman and Malaysian arcs—like the basement folds of the Turan did with respect to the Iranian arcs.

The function of transverse alignment that distinguishes the 63rd meridian was fulfilled from the extreme south of China to south of Java, by the 110th meridian. The flow lines with a general direction to the south along that transverse alignment—the present-day direction—have undergone at the right wing gradual deviations to the southwest and west; the flow lines at the left wing have undergone symmetrical deviations to the southeast and east with the formation, in what was going to become the Philippines, of a virgation of the second type (confined virgation), which today closes up northward in Luzon and opens to the southwest. But the occurrence of five marginal seas inside the protuberance, and of two oceans along the periphery, leads us to think that these inferences are true only with reservations of time and space, to say nothing of subsequent deformations according to another style.

In essence, Angara Land has launched two flank attacks—or two flank counterattacks—depending upon whether one attributes the greatest displacement to it or to India. During this time, the Angaran center was depressed and India was suffering less, although the great complex basement fold of the Himalayan zone, with its clean-cut thrusts, is one of the obvious results of this situation.

The southern part of China gradually yields its secrets, but it would be premature, in spite of all that science owes to L. Loczy, F. Richthofen, A. Leclère, and B. Willis and in spite of two good recent attempts at a cartographic synthesis, to give a somewhat integrated sketch of it. Any attempt of this kind would rapidly become obsolete. The recent creation of a Geological Survey of China, which has already given to science several works of high quality, forecasts very clearly the immense future promised to our Chinese colleagues.

Therefore, I shall limit myself to a few scattered glimpses. The Precambrian and the movements of the three subsequent cycles are certainly represented. The dominant direction is northeast. (287)

A first complex covers the southeastern part of the land; to the north, it extends a little beyond the line of the Yangtze; its major part, located east of the 112th meridian, projects, however, strong spurs—warped anticlinal structures or voussoirs of basement folds—to the southwest in the Kwangsi. This first complex includes crystalline schists, folded Paleozoic including Devonian and Dinantian, folded Permian, Triassic, and Jurassic Angara beds, granite, and a variety of volcanic rocks. The role played by the Alpine cycle is important; granites have sometimes pierced the Jurassic. The role of the Hercynian cycle is very probable. A bundle of folded Paleozoic that as a whole follows the bend of the Yangtze from the vicinity of Nanking to that of Hankow displays a concavity to the north on the 116th meridian, near Chinchiang; the arrangement of this area seems to announce the resistance of a promontory—oriented to the south—of the great massif of gneisses, probably very old, that extends to the common boundaries of Anhwei, Honan, and Hopeh.

The Anthracolitic, Triassic, and Jurassic covers that spread out with folds and with a predominantly tabular character over most of Kweichow, over large areas of the west of Hunan and of the southwest of Hopeh, obviously conceal a very rigid massif that could well be the southwestern extension of the former and build with it a second distinct region. A large spur of ancient formations extends southwest of Tungting Lake and disappears in that direction under the above-mentioned sedimentary cover. The ancient massif, intersected by the Yangtze upstream of Ichang, with its Precambrian, its envelopes, and its superficial intersections showing concentric curves, is a beautiful Alpine basement fold; its axial dipping, southwest of the river, is rather steep in that direction, but it reappears farther on in a new axial culmination, in the area of the 30th parallel and the 110th meridian.

The folds consisting of Paleozoic and Angara beds in the northern Yunnan and the basin of Szechwan have very poorly known relationships. They indicate, beyond the second region with its large tabular structures, the reappearance of more folded conformations. In the eastern and southeastern parts of the Szechwan basin, the Alpine folds are pressed together when nearing the second region and display oblique overlaps as in the vicinity of a resistance. This situation seems to confirm the idea of a particularly rigid massif beneath the gentle coverings of the second region.

A long belt of Alpine foldings, which are sometimes of cover type but more generally of basement type with reworking of ante-Alpine heterogeneous older basements, thus surrounds the ancient nucleus, which is probably Precambrian and is located in the center of the second region: in the southeast, we find the folded elements of the first complex; in the south and southwest, the thrusted arc of northern Annam, Tonkin, and southern Yunnan; in the west, the bundle of northern Yunnan; in the northwest, the folds of the Szechwan basin, a portion of which is perhaps of basement type) (288) in the north, the Chinling Shan. The second region itself shows basement folds and cover folds.

East of the terminator of central Asia, the basement folds of the Kunlun, having escaped from the great compression, plunge rapidly in a southerly direction, toward Indochina; in a southeasterly direction; and in an easterly direction into the Chinling Shan. Their still powerful masses dominate and surround the margins of the Szechwan basin in the west, northwest, and north.

The Chinling Shan is a large Alpine basement fold, a reactivation generated from the north by the southern margin of the Sinian massif, which is not very far away. The cross-sections of W. A. Obrutchev reveal, for the territory enclosed by the 104th and the 106th meridian, and the 32nd and the 35th parallel, an old basement, mainly of Hercynian age, with folds often overturned to the south, under which appears in some places a pre-Devonian substratum. The whole has been warped into a large Alpine basement fold with unconformable covers of Upper Carboniferous sandstones, continental Jurassic, and Gobi beds that often display gentle synclinal undulations subordinated to the large basement anticline. The best developed of these synclines shows, in the Gobi beds, a shoving to the north. Farther east, the Chinling Shan displays along its southern margin Alpine foldings whose age is, according to B. Willis, later than some insufficiently dated horizons of the Lower Mesozoic. Granites pierce this basement, and the whole is overlain unconformably by younger Angara beds. It is quite possible that the old massif of southern China may have contributed, along more than one transverse alignment, to initiating the reactivation of the Chinling Shan.

In the Kansu, where between the 104th and the 107th meridian the deviated branches of the eastern Nan Shan terminate, and perhaps already in part on the Sinian massif, one finds folded beds, supposedly of Pliocene age, extending over large surfaces and overlain by a thick mantle of loess; the substratum appears only here and there. These cover structures, often very intense, are certainly related to the recent phases of the deformation of the Nan Shan and perhaps of the Chinling Shan.

It is not necessary to mention the more northern regions of the continental mass from the Chinling Shan to the Sea of Okhotsk; I have previously described them in as much detail as was useful in the present state of their exploration.

The Aleutian festoon, the Kamchatka, the Japanese festoons, and those that follow each other from the Philippines to Burma present, as does all of East Asia, the fundamental problems of the nature of the bottom of the ocean and of the marginal seas and of the influence this nature exerts, as we can well imagine, on the behavior of the visible and invisible structures: the problem, in essence, of the fixism or of the unlimited mobilism of the continents.

XXVII. Concrete Tectonics and Theories of the Earth

Until now, a steady desire for concrete tectonics has led me (289) to search, without paying much attention to the theories of the Earth, for the solution of the ancient problems as well as the answer to those that appear at every step. During this process, I have discovered many features, and among them, results of general implication that are bound to direct my path during the brief but inevitable confrontation ahead of me. The reduction of vertical movements to the rank of small effects subordinated to an essentially horizontal plastic flow; the connection of axial behaviors to events of the same order; the high degree of relationship that appears in so many continuous aspects; the certitude that continental slopes overlain by new and very plastic deposits can behave approximately as a half of a geosyncline and give rise to new chains of important tonnage; the ponderous and powerful behavior of basement folds; the easiness with which plastic deformation can involve the entire mass of continents in spite of their high degree of induration; and the complete reversal of the accepted ideas pertaining to the energetic preponderance of geosynclinal chains: such are the results whose impact will be difficult to ignore.

The same desire for a concrete approach has led me to present the discussion on the interior of continents almost as if the debated problems did not exist on a higher scale.

In regard to certain marine boundaries of Eurasia, this approach could not be maintained rigorously, as we could well see. In East Asia, such an approach becomes entirely impossible unless one limits oneself to a mere analytical catalogue of the known facts.

We have theories by the dozen, but their very number decreases the chance of agreement among them. Today, it seems that the dispute is focused on the theories that imply the fixism of continents and the hypothesis of large-scale drifting as visualized and powerfully presented by A. Wegener.[26]

A complete confrontation of these theories among themselves and with the facts shall not be attempted here. Even limited to the requirements of geology, or simply of tectonics, it would exceed my present purpose; I may perhaps present my reasons somewhere else. However, I have undertaken this confrontation for the entire Earth within the limits of my means. Since 1915, and particularly since 1918, I have thoroughly scrutinized the theory of continental drift and its application to the entire spectrum of tectonic forms available to me, and to all the types of movements I can visualize. Therefore, if time is lacking today to justify some of my interpretations, one should not, in all fairness, consider them as hasty or devoid of foundation.

For the sake of abbreviation only, I shall use the expressions of fixism and mobilism[27] to characterize the essential of the two attitudes.

Fixism is not a theory but a negative element common to several theories. All things being well considered, it is essentially the absence of position (290) versus a problem that is precisely that of mobilism, and it can be defined only with respect to the latter. Strictly speaking, it can be neither demonstrated nor refuted; such is the fate of any idea that relies on the absence of testimonies.

The constructive power and the dangers of fixism begin only by its association with positive views, to which fixism communicates the principle of inertia, which is intrinsic to it. Fixism is usually, but not necessarily, associated with the theory of contraction and with the concept according to which the oceanic substratum is of same nature and of same thickness as the continental substratum, from which it is supposed to differ only through the effect of large-scale collapsing processes of radial orientation.

In agreement with many of my contemporaries, I consider this interpretation as totally incompatible with isostasy. Therefore, I could now stop this discussion of traditional fixism and proceed immediately to other considerations. However, I would like to indicate the aim to which it seems to point and the final picture it provides of things.

A patchwork of continental areas and of geosynclines, of rigid portions and of flexible articulations, extends under the oceans as well as under the continents and their adjacent areas. Whether this patchwork remained upright or sank through large-scale collapsing or was not exalted does not really matter in a first approximation; the great rules of the behavior are everywhere the same. Geosynclines are formed; they are filled with folds that tend to be arranged in double chains and sometimes differently; these geosynclines form anew while changing position; and the process repeats itself until extinction of the power. Large-scale collapsings take place which generate lateral compressions. The chains react by foldings and thrustings, which even if of considerable magnitude remain nevertheless local with respect to the whole. By looking at this entire picture, one sees that everything becomes *encrusted* in place, with time, in the vertical sense.

This classical concept, assuming it is correct, should naturally agree with the largest features revealed by our investigation and particularly with the behavior of the basement folds, the main expression of folding on this planet and a reality independent of any theory. This concept becomes easily complicated by the basement folds that develop within all the continental masses and particularly, through reactivation, in the indurated ex-geosynclines. Leaving aside any pretense of theoretical explanation and restricting oneself to an imagery of movements, one would conclude that there is nothing here that cannot be reconciled.

But, the difficulties of the theory of contraction—assuming it is related to fixism—are greatly increased by the necessity of accounting for the basement folds. Some have already considered contraction as inadequate to explain the total shortening due to ordinary foldings. A lot could be said on this question and even more on the preliminary question that pertains to the means of estimating the shortening. But, I shall not insist. Basement folds display, as do new chains, large thrusts that do not facilitate either the estimation of the shortening or (291) the task of fixist contractionism. But they imply, in particular, an energetic expense that surpasses by far anything one has wished to consider until today. To estimate the tonnage of the slightly ordered basement folds for the entire Earth to be about ten times that of the new chains is an assumption that remains certainly well below reality. Since the great old shields are essentially very broad basement folds, large *basement brachyanticlines,* one should say fifty or sixty times or even more. And after all, this is only the volumetric ratio; in order to proceed to the energetic ratio, one has to take into account the high degree of induration and multiply everything by a coefficient that cannot be precisely given today but that certainly should have a high value. These are indeed numerous subjects for meditation, presented in a few words.

The classical concept combined with that of basement folds certainly allows bold interpretations. Here is one that I have considered: the Mediterranean-type seas, the marginal seas, and the oceans are but basement synclines. These geosynclines of a new type, formed by lateral compression and becoming the location of more particular types of lateral compressions, generating chains, would unquestionably explain many features. In that respect one thinks immediately of all kinds of island festoons, of the Oceanides, and of the elongated crests that sinuate in the middle of the Atlantic and in the western portion of the Indian Ocean. This concept leads directly to the idea of the continuity and particularly of the universality of folding, which becomes the only major aspect. Indeed, considering from this viewpoint the closed environment formed by the entire planet, one encompasses in one swoop, and rightly so, the totality of the horizontal and vertical aspects of the deformation. It becomes completely useless to ask oneself if the radial movements follow or precede originally the tangential movements, and what their reciprocal relationships are. This question, debated by generations of geologists, is justified on the scale of small entities but is meaningless with respect to the whole. The incapacity of the plastic media to transmit, beyond a certain distance, an effective effort is not an insurmountable obstacle if one assumes for the upper part of the oceanic substratum the same kind of heterogeneity that is displayed so clearly by the continental substratum. Thus renewed, the classical concept would allow extensive enrichments, and a long time would elapse before these resources would be depleted. Unfortunately, in relation to all this, there is isostasy, and as we shall see, much more.

The fixist hypotheses admit sometimes the possibility of small horizontal movements. The center of shape of a continent would thus complete small displacements around an average location. But, in such a case, where should one stop along the path of drifting?

I have pondered at length the tectonic consequences of the hypothesis according to which the oceanic substratum would consist of sima completely overlain by a sial thinner than that of the continents. This hypothesis, (292) whose principle goes back to G. B. Airy, has been developed by O. Fisher and G. Lippmann. It allows us to visualize continental drifts hindered by the film of sial remaining at the bottom of the ocean, which this push helps fold. Applied to the Oceanides and to the objects of similar aspect displayed by the Indian Ocean and the Atlantic Ocean, it would account for the elegant freedom with which these arcs seem to be distributed. One would deal with basement folds of a particular kind, with *folds of a thin basement*, that would behave with a certain agility—covers included—and that nothing would prevent, under appropriate circumstances, from becoming complicated by new chains. One can often succeed in making the hypothesis of the thin basements qualitatively compatible with the distribution of gravity; yet this hypothesis extended to the entire oceanic domain is considered to be in little agreement with the distribution of the terrestrial magnetism. Therefore, one should accept, with a certain reservation, the idea of the above-mentioned deformations. However, it should be remembered that if the theory of large-scale drift were true, there may in addition to a basement exist remains of sial thinned by previous stretchings in which such deformations can take place.

The validity of a theory is nothing else but its capability of *accounting* for all the known facts at the time it is presented. In that respect, the theory of large-scale continental drift is of flourishing validity. In its incipient stages, it was aiming at the absolute; subsequently it gained a lot in strength and in flexibility without sacrificing anything of its rational structure; on the contrary, it became enriched and increasingly in harmony with the vision that leads the whole. This work of clarification and of improvement is very obvious throughout the sequence of works by A. Wegener. Strongly documented at the meeting points of geophysics, geology, biogeography, and paleoclimatology, it has not been refuted. One has to have searched at length for objections, and particularly to have found a few, in order to estimate properly the kind of immunity that distinguishes it and that originates from a great flexibility combined with a great richness in operational possibilities. One thinks one has found a decisive objection and that after another one the whole theory will collapse. In fact, nothing of that sort happens; one has only forgotten one or several mechanisms. It is the protean resistance of a plastic universe.

Certainly, there are numerous objections, but almost all of them belong to the above-described type. Among those that have been published or of which one can think, only a few are valid. These pertain to accessory aspects and never, in the present state of our knowledge, to vital aspects of the theory.

Therefore, it is quite true to speak of the validity of this theory in the sense mentioned previously.

The nonexistence of a refutation should not, strictly speaking, be considered as a proof. Still strictly speaking, positive testimonies may have only the value of arguments. New facts may be discovered that have the strength of an unyielding obstacle. Other theories, (293) either fixist or mobilist, may be born, or reborn under rejuvenated forms. Did not M. Bertrand propose, with the impetuosity of his last works, to rotate as a whole an envelope of the globe compelled to deform over a core?

But presently there seems to be no fact sufficiently contradictory to prevent us from enjoying, on those rafts on which our fates float, the delight of the sensible transports to which A. Wegener invites us.

XXVIII. Problem of the Pacific Ocean, Fixism, Mobilism

The value of a theory lies entirely in the agreement between the consequences derived from it and the well-observed facts. It lasts just as long as this agreement, of which the present mobilism displays numerous examples, which I do not have to recall: I am not giving a lecture. However, it is useful to make a few new soundings before embarking upon new adventures.

I am planning to disclose by statistical criteria some of the more general conditions that prevail under the Pacific Ocean and that therefore stand a chance also of prevailing without appreciable changes under the other oceans. We cannot see anything of the substratum of the Great Ocean, but we cannot doubt that the behavior of the Circumpacific chains depends to a great extent on the nature of this substratum. These chains present, by their tonnage and its fluctuations by segments, an approximate means to estimate their energetic quota, their endowment in useful energy—that is, in energy consumed by the internal deformation including its clean-cut effects. One should consider only the largest arrangements and the greatest durations so as to eliminate, by the law of large numbers, the influence of the small fluctuations, either local or temporary. Submitting to the same treatment the chains of the Tethys and their framework of basement folds, one has the means to compare the quotas belonging to this half-belt of the globe, to the Pacific belt, and to the great segments that build both of them.

From the numerous inferences that can be made from such a procedure, I shall discuss only those that are more pertinent to this debate.

One has generally the means to estimate separately and with an adequate approximation, for a given segment, the *reactivated* tonnage of the basement folds and the *new* tonnage belonging to the new chains. The ratio of the two tonnages is useful to consider in many cases. Their summation is the *total* tonnage of the segment. In order to consider the energetic quota, it would be necessary, strictly speaking, to give to the new tonnage an average coefficient appreciably lower than that attributed to the reactivated tonnage. Since the new tonnage rarely represents more than a minute fraction of the reactivated tonnage, one would not commit a very large error in considering the energetic quota as proportional (294) to the total tonnage, as long as very large objects are considered whose deformation has encompassed a very long span of time. In order to make the results comparable from one segment to another, one should divide the total tonnages by the length of the corresponding segment. Thus, one obtains the *normative* tonnage, a tonnage per unit length. The fluctuations of this value shall give, except in special cases, an approximate picture of the general behavior of the external resistance.

Examination of the most significant facts shows that they can be associated in three orders.

First, the normative tonnage displays greater fluctuations in the half-belt than in the belt. At the lower limit, both arrangements start with low values; at the upper limit, no Circumpacific segment equals the segment of central Asia. The most vigorous among them, the American segment, extending between the 34th and the 49th parallel north, remains noticeably lower.

The tonnage of the Oceanides is certainly considerable, but it is spread out over such large areas that one cannot attribute to all its parts, in regard to Australia, the function of peripheral chains, which is so obvious in the remaining portion of the Circumpacific arrangement in regard to Asia or to the two Americas. One could not do so, at least not without a biased hypothesis. It is necessary, in this kind of consideration, to substract at least all the tonnage of Micronesia and of Polynesia, whose chains do not have the appearance of being centered on Australia as do the arcs of Melanesia, Fiji, Tonga, New Zealand, and the more internal regions, an appearance that discloses, at least for a portion of them, intimate genetic and dynamic relationships with that continent. In its normative tonnage, the remaining portion of the arrangement—the internal Oceanides, which are really Circumpacific and not Intrapacific—is below the segment of central Asia.

The lower degree of amplitude of the fluctuations discloses, for the whole of the Circumpacific chains, a less unequal behavior, a behavior less subject to extremes than that of the half-belt.

Second, the normative tonnage is appreciably higher for the whole of North America than for the whole of East Asia. The endowment of useful energy that is consumed in foldings has been appreciably lower in the latter case.

Third, this inferiority of East Asia is increased by the fact that the new tonnage, less exacting, is greater than it is in North America, both in a relative and an absolute sense. This latter continent displays almost no new tonnage. After having subtracted the sierras of eastern Mexico, already of Mediterranean type, and the new parts properly Alpine of the Coast Ranges in a broad sense, we are left with only Laramide basement folds consisting of very ancient dead material and Alpine basement folds made of Andean material. Powerful basement folds certainly are involved in the new chains (295) of East Asia, as can been seen particularly in the great islands of Japan whose reactivated tonnage is probably greater than the new tonnage. This can be noticed also, but to a lesser degree, perhaps because of a lesser average exaltation, in Sumatra and in other parts of Indonesia. But with respect to new tonnage per unit length East Asia as a whole is unquestionably superior to North America.

Any theory about the Pacific should account for these facts. I would like first to point out, without prejudging fixism or mobilism, that *the strong fluctuations of the deformation of the Tethys, including its rear parts, disclose the predominant influence of the heterogeneity of the two sials confronting each other, of their complicated conformation, and of their hard collision.* Vice versa, *the weak fluctuations of the Circumpacific tonnage disclose the presence, or predominance, beneath the Great Ocean, of a more homogeneous and more yielding medium than the heterogeneous and always very resistant higher parts of the continental blocks.* Therefore, there is no collapsed Pacific continent and, consequently, no Circumpacific geosyncline. The chains of the belt, insofar as they consist of basement folds, arose from the continental basement or slope. As to the new tonnage, it includes the folded covers, continental or neritic of the continental basement, and the chains that arise from the freshly deposited sediments, either neritic or bathyal. These sediments are accumulated in great thicknesses on the continental slope and sometimes at its foot, as a consequence of submarine slumpings that are responsible at great depths for the repeated intercalations and the mixture of these deposits with the abyssal sediments. Therefore, the abyssal sediments may be brought to the surface in small quantities. The behavior of the slumpings is the same as that occurring on the two opposite taluses of a geosyncline or on the two opposite slopes of a submerged geanticlinal cordillera. The single slope can, during the evolution of these chains, become subdivided through folding into cordilleras and furrows as would happen to the two slopes of a geosyncline. The deepest of these furrows may contain abyssal deposits, again as in a geosyncline. The weight of the thicker deposits, either bathyal or neritic, that are accumulated on the main part of the slope can express itself after a long time by a very broad furrow of isostatic origin, which depresses the entire overburdened zone without necessarily modifying, to an appreciable extent, the shape of the surface of the slope. However, this large alveole, which is not limited on one side by any continent, should not without exaggeration be called a geosyncline. The same should be said about the entire system. Circumpacific basement folds, new folds, and new chains result from the deformation of the continental margin and in particular of the slope. It is from the highly deformed slope that rise the basement folds that appear in the very center of new chains, as in Japan, and in some parts of Indonesia. The bulk of the new chains also originates from the slope but from its upper portion consisting of easily deformable sediments. In order not to confuse these new chains (296) with real geosynclinal chains, I shall use the term "marginal chains," which emphasizes the location of these objects on the threshold, at the entrance of the continental domain, and which implies, with an initial *monoclinal* arrangement, the deformations just discussed.

The present-day mobilistic theory accounts, without difficulty, for the facts pertaining to the distribution of the tonnage and for their immediate interpretation. According to this theory, the relatively homogeneous and yielding medium that occupies the Pacific is the sima. Of course, the behavior of the sima with respect to the sial can present great differences according to the various manners in which the plasticity or viscosity of the media and the duration and intensity of the efforts intervene. But, for continental drift to occur the deformations in which the sima behaves as a yielding medium must predominate in the course of time over those in which the sima resists and compels the folding of the sial. From a predominantly statistical viewpoint that encompasses the entire globe and the whole duration of the deformation, it cannot be otherwise if there is a sima over which large rafts of sial are drifting.

The mobilistic theory explains easily the second and the third group of facts, in which the energetic inferiority of East Asia versus North America is expressed. It admits bow stresses that compress and fold the sial against the sima, under certain circumstances. It also admits stern stresses[28] that consist of a retraction of the sial from which results, for that material, the more or less complete interruption of folding with the predominance of traction effects: distensional fractures, buttonholelike tearings creating marginal seas, releases of cordilleras that from then on lag behind the continent in the form of somewhat severed festoons. The sima, on the other hand, is compelled to adapt itself to so many new conditions, and it rises under the marginal seas and in the spaces recently abandoned by the festoons, on the stern side. Because of the delay with which this rising occurs in the end, deep furrows result, which according to the classical concept are considered to be foredeeps. Since the mobilistic theory requires that bow stresses should have been predominant along the western margin of America, and the stern stresses should have lasted for an appreciable length of time in East Asia, the superiority of the former and the inferiority of the second with respect to folded tonnage are self-evident.

The elegance with which the theory of continental drift explains these fundamental facts, unknown at the time it was conceived, is certainly a great argument in its favor. None of these facts rigorously demonstrates the mobilistic theory or simply the hypothesis of the existence of the sima, but all of them are perfectly in agreement with both, to the extent of making them very likely to be true.

The fixist contractionism with collapsed continents can attempt to explain the same facts, but this entails two risks. For the explanation of the fluctuations, it will require the intervention of changes of power (297) rather than those of resistance in order to set aside as much as possible the idea of an essential difference in nature between oceanic substratum and continental substratum. But the extraordinary degree of filling, of synergy that is inherent in this initial assumption of fixist contractionism, implies a certain equality in the original distribution of power and requires the explanation of the segmentary differences, mainly through resistance: hence, a beginning of internal contradiction about which one cannot say yet to what extent it undermines the economy of the system. The second weak point is the fortuitous character taken on, in this hypothesis, by the fluctuations of the tonnage and of the energetic quota. The facts reveal a statistical law that emerges from randomness and dominates it; mobilism agrees in advance with this viewpoint; qualified fixism must attribute everything to randomness. This disagreement is profound because any statistical law has a physical reason in the inorganic world.

If one considers the normative tonnage of Andean age by itself, one finds it more important in America than anywhere else. It predominates in New Zealand, which is very natural since Australia, still connected to South America through Antarctica, played in the past the role of a bow as required by mobilism.

In East Asia, the Andean tonnage is still rather considerable but deficient with respect to North America. The original arrangement as a whole being the same, it appears that for East Asia this difference stems from the intervention of stern stresses of Andean age, subsequently erased by Alpine bow stresses—in particular during the paroxysms—and finally reactivated, at a recent time, with the amplitude one can presently observe. There is not only succession in time but, in addition, alternation of the two types of stress. This concept should be kept in mind whenever complex cases are to be unraveled. Besides, it is probably more general than that of a simple succession; therefore, it has a greater chance to apply to a variable degree of geosynclines and monoclines as a whole. Consequently, the total drift would express itself, for a given material point, by rather complex trajectories that would form interwoven curves and sometimes broken lines recalling a large-scale Brownian movement; with the passing of time, a westward component and another toward the equator would predominate. Because of these complicated trajectories, the deformations of monoclines into marginal chains, and the deformations of geosynclines, can undergo a thousand kinds of oblique and longitudinal distortions: it is a considerable enrichment of the imagery. The same circumstances explain also, by means of internal frictions of the sial, the slow changes of azimuth of the flow lines, over time, within the intracontinental fluxes.

The mobilistic theory accounts at least qualitatively for basement folding. The raft of sial is deformed and folded while drifting. The frontal resistance of the sima, along the margins, is efficient under certain conditions; its basal friction along the lower surface of the sial, and the more complex frictions (298) that develop in the arrangements in which the two media interpenetrate each other or grade into each other, lead also—favoring the deformation of the sial and the basement folding—to a withdrawal from the energy of drifting. One can visualize still other kinds of derivation of energy. The internal frictions of the sial take care of the rest through gradual spreading. The deep-seated portions of the sial certainly undergo a very active and unquestionably rather equally distributed deformation of the matter. The basement itself, that is, the base of the upper part of the sial, namely, the substratum of the heterogeneous patchwork, of the *tectonized zone,* introduces the specializations of the flux that we have investigated and localizes the basement folds, which in turn transmit their warped shape to the top of the upper part of the sial—subsurface and surface—with all the modifications and fractures that result from the final propagation of the energy within these less flexible media. At the very base, large lenses of sial, the *deep-seated lenses,* can be generated, by isostasy, under the main projecting features of the surface, as shown by Albert Heim for the new chains and particularly for the Alps.[29] These lenses often have the tendency to spread over a width greater than that of the visible chain.

It seems as if basement folding marked a few but missed some additional behaviors at the places at which the sial is thin.

Naturally, the mobilistic theory will have to find an adequate source of energy not only for the work of drifting but also for the great work required by the development of basement folds.

It is almost superfluous to recall the traction processes assumed by the mobilistic theory. They are in a sense the complementary phenomenon of basement folding, the result of a negative compression. Large-scale fractures, not always easily distinguishable from those generated, in a secondary manner, by basement folding; ripped buttonholelike depressions in which the sima will eventually appear; formation through continued separation of seas and oceanic spaces; and finally festoons, kept in leash or set completely free—all are considered the expression of traction processes. The origin of the latter can be found for the most part in the variations of the basal or frontal adherence between sial and sima; in the mechanical regime of the *sponges* through which the two media interpenetrate each other; and, finally, in the *anchoring* that certain deep-seated portions of the sial can display in a transitory manner.

In the folded rings of the Circumpacific arrangement, containing basement folds and marginal chains, the distribution of the useful energy cannot be in all aspects identical to that in the geosynclines and in their rear parts with basement folds. In both types of arrangement, the basement folds are generated mainly by the intracontinental energy. The energy of compression, of considerable magnitude in the first arrangement while the sima resists, takes on less importance or is missing while the sima yields. With respect to the energy restituted by the marginal chains, its part may be appreciable (299) when the overturning of these chains takes place toward the continent; it may decrease to a small amount or may be missing when the overturning occurs in an oceanward direction. In both arrangements, the withdrawal earmarked for the new chains is mainly provided by the intracontinental energy.

The mobilistic theory has somewhat neglected the concept of geosyncline. It is therefore appropriate to sketch a connecting link. A geosyncline will generally result from a horizontal *traction* that stretches the raft of sial. The stretching is at first easier in the deeper part of the sial rather than in the upper part, where extension fractures may develop. While thinning, the sial sinks and develops a depression: the subsidence inherent in the geosynclinal process does not, therefore, stem from an original radial stress; it is only the vertical effect of a horizontal distension. The overburden of the deposits helps of course to accentuate the alveole, but the latter is not necessarily the original feature. Until compensation, the sima rises under the thinned sial; this behavior accounts for the frequent association of green rocks with bathyal and abyssal sediments. The mixture of abyssal with shallower sediments takes place through submarine sliding on the slope. The vertical margins of the thinned zone, which have preserved the normal thickness of the sial, represent the jaws of the geosyncline. Whenever compression replaces traction, the jaws are brought closer and the classical geosynclinal deformation begins with its embryonic folding by means of cordilleras, furrows, and true foredeeps: the conclusion is almost always the formation of two geosynclinal chains with opposite overfolding.

If traction continues, instead of giving way to compression the sial continues to stretch and the sima appears at the bottom of the alveole. Along the transverse alignments where such a situation occurs, the geosynclinal condition is replaced by the oceanic condition; if such a situation becomes generalized, only an ocean is left. If a compression occurs at this stage or just before it, when the sial is really very thin, the lack of synergy will lead to the generation of one or two trends of marginal chains, of Circumpacific type, and not of the double chain of geosynclinal type. If the compression continues, the latter type will establish itself gradually and may perhaps eventually predominate, but the traces of the simple or double marginal condition will persist, although veiled.

The ordinary geosynclinal behavior is therefore, in essence, an incomplete lenticular cutting up of the continent; whenever the cutting is complete, the oceanic condition appears.

In addition, events can occur in a more brutal fashion and proceed by means of high and steep distensional fractures that eliminate the stretchings and result in more massive jaws—hence the differences in the subsequent flexible deformations when they exist.

All these geosynclinal deformations may become complicated in the same manner as the corresponding marginal deformations, namely, by the effect of bow and stern stresses, (300) of oblique and longitudinal behaviors, in reciprocal alternations or in simple succession. The resulting complexity has an appreciable chance to be greater in a geosyncline than in a double marginal arrangement because of an effect of synergy.

Small intracontinental geosynclines can be generated without stretching, through the simple continued effect of an overburden of sediments. As long as the original factors operate without outside interference, these alveoles remain almost full while subsidence operates along the substratum of their axial zone.

There are between all these types of geosynclines transitions and interactions that I shall not discuss here.

In the case of subsequent drift, the continent can glide over deep-seated lenses that in plan view will find themselves out of line with respect to the visible intumescences. These lenses will lag, variably deformed or spread out, behind their original position, generating rising effects in their new location, which because of this import of sial generally displays appreciable negative anomalies. A. Wegener has shown this behavior for the under-Alpine lens, which a drift of the continent to the southwest has set out of line and left dragging in the northeast, while pushing it partly under the southern regions of extra-Alpine central Europe.[30] Portions of lenses, perhaps entire lenses, can remain behind as a result of similar behaviors. On the bow side, the advance of the margin of the chain, with respect to the margin of the lens, may go so far as to lead to the appearance of positive anomalies with a more or less zoned distribution. In the rear, the destruction of the lenses can be, in the case of important drifting, carried to the complete release of the sial that formed them; this material, hereafter separated from the continental mass and highly deformed, is supposed to rise slowly within the sima, on the stern side. The different behaviors just described represent the process of *drift over lenses,* which displays different degrees and an evolution of which the complete release is at the same time the final term as well as the prelude to other effects.

It is obvious that the drift over lenses cannot be considered as an accessory effect: it is a fundamental aspect of continental drift. The shapes previously acquired by the folded elements—during the original axial behaviors—are modified by the isostatic risings and sinkings that the drifts over lenses, by reworking the base of the sial, are bound to generate. These modifications are added to those created by the behaviors that we have called accessory. But the fact that the rule of axial behaviors is verified statistically at present as the global and average result of all the flexible deformations that took place during very long spans of time shows that neither the drifts over lenses, regardless of the importance of their effects, nor the accessory behaviors, nor the association of these two kinds of perturbations have led to a general destruction or simply to a very important reworking of the visible conformations. The reason (301) for this is very obvious. Drift over lenses, like the secondary behaviors but probably more efficiently, tends indeed to destroy the previous conformations, either visible or deep-seated. But, at the same time, the effort of drifting tends to maintain, to accentuate, to complicate the major part of previous foldings, to enrich it with entirely new folded elements, including all the features that these actions imply with respect to constructive infratectonic features, to increased lenses, and to new lenses. Therefore, a battle goes on between the destructive alterations, on one side, and the restorations, the constructive reworkings, and the new constructions, on the other side. One can imagine that the first group of factors may sometimes predominate as a fluctuation of an average statistical condition for small objects considered during short periods of time, But, in regard to the whole, which we are now attempting to embrace, to the large territories worked and reworked over great lengths of time, the predominance belongs, nevertheless, to the second group, namely, to the constructive behaviors. Some sort of compromise occurs between the two kinds of processes and things are from the start in favor, or gradually changing in favor, of folding and its infratectonic consequences. Our rule of axial behaviors shows this compromise on the scale of the new chains and of the basement folds, in the ordering of the visible and the concrete.

On the continental scale, the slow uplifting that the entire sial undergoes is, according to A. Wegener, the result of repeated compression and folding, expressed in a conceptual manner by a contraction of the hypsographic curve. This view on the continental scale is comparable to my view of axial behaviors on the scale of the new chains and of the basement folds. In both cases, one deals with objects that become folded while undergoing compression and rise while being folded, if only the predominant behavior is taken into consideration. Therefore, it is in a sense permissible to consider each continent as a very large basement fold—removable and divisible—to whose behavior a great number of more particular deformations are subordinated: basement folds shaped as shields, domes, or platforms; well-ordered basement folds; traction effects; new geosynclinal or marginal chains; infratectonic effects; cover foldings; and so on. One can say of this huge comprehensive basement fold that its radii of curvature are extremely large and that isostasy, particularly effective on this very large scale, imposes upon the carapace of the fold one of the smoothest, most regular, and most general conformations, combined with one of the most well determined rates of rising.

The agreement between the two viewpoints, on the scale of the basement folds or of the new chains and on the continental scale, is so much the more remarkable that it was the less searched for. Originally, the concept of axial behaviors came to my mind through the relationships existing between the axial culminations and depressions of the Alpine folds, on the one side, and the obstacle massifs, on the other. I have touched upon this subject many times, both graphically and in writing—my earliest publication is dated 1911.[31] (302)

One can perceive in the sketching or the recalling of so many types of behaviors the complication of what is meant by plastic flux, this essentially horizontal flow to which are subordinated, both in the sima and in the sial, thousands of effects of which one can only guess the smallest part today.

The most noticeable exceptions to the rule of axial behaviors by subsequent reworkings correspond to fragments that are separated by distension from the great continental masses under conditions that prevent a renewal of the folding or that profoundly modify its process. The more or less released festoons have a chance to belong to this kind of exception, and the axes of their folds to display twenty kinds of readjustments, a few of them active in some manner in a folding that is modified in its conditions only, the others passive through simple isostatic adaptation to the substratum presented by the sima.

Essential differences exist between my basement folds and the folds with great radius of curvature *(Grossfalten)* discussed by E. C. Abendanon.[32] My basement folds having been defined, their major shapes described, and their most important behaviors illustrated, these differences are self-evident.

XXIX. Glances at the Atlantic

In the synthesis presented here, very general facts of unexpected implication have been disclosed. I have sketched an imagery of movements that agrees with that of mobilism, while extending this concept; this imagery owes also a great deal to the observed tectonics, which it similarly extends. I have thus assembled a collection of compatible mechanical artifices with which I am going to reexamine, without intending to do more than sample a few behaviors, regions I have superficially discussed before. Regardless of the kind of object, whether oceans or ancient shields, basement folds or marginal seas, marginal chains or geosynclinal chains, I shall examine them with only one preconceived idea, namely, the assumption that the combinations of the most complicated behaviors are the most probable; yet I shall keep in mind the possible reduction of their number in light of facts that would allow me to do so. The reservations that I have expressed with respect to any nonconcrete tectonics shall follow me as a silent background in the narrative.

The *North American continent* shows the process of drift over lenses on a grandiose scale. This interpretation is immediately suggested by the distribution of the Bouguer anomalies in the United States. The front on the Pacific coast shows positive regions about which one cannot say as yet to what extent the default of gravity due to drift over lenses, the positive perturbation of Helmert related to the upper crest of the continental slope, and the more local distribution of the densities should participate in the explanation of the observed anomalies. But in the rear, (303) where the drift over lenses is very obviously displayed, there is evidence that it would be dangerous to eliminate this factor from the explanations pertaining to the frontal part. The zone of strong negative anomalies extends hundreds of kilometers to the east of the visible Laramide front, beneath the High Plains. In the same direction, the major portion of the anomalies gradually decreases and the lens similarly thins while its lower surface rises and the sima with it. The major part of this conformation ends along a winding line—the general trend being meridian—enclosed between the 103rd meridian to the west and the 97th meridian to the east, with some reservations due to the degree of approximation compatible with the large spacing of the stations. Possibly, a solidary compensation, involving both the region of the chains and that of the High Plains, may lead to a sinking of the latter, thus contributing to its lack of mass. But since nothing similar occurs on the west side of the chains, the unilateral character of the situation is obvious, and I can visualize only drift over lenses accounting for the essentials of it.

Farther east, beneath the center of the United States, weak negative anomalies seem to prevail with noticeable positive exceptions, which I shall not discuss. Farther away, the sub-Appalachian lens appears with respect to the visible intumescence slightly off the axial line, toward the east or southeast. The effect is more moderate than it is for the lens of the great chains of the west, but it is of the same direction, and drift over lenses explains the essentials of both conformations. While drifting to the west, the continent tends to abandon a portion of its deep-seated lenses, which remain behind.

My *Proto-Atlantic,* a Caledonian geosyncline that in some parts is much older and whose traces run from the Arctic domain to the Antarctic regions, is perhaps not irreconcilable, for all we know, with the fixist imagery. The large-scale collapse processes assumed to have generated the present Atlantic would have renewed the general arrangement of the ancient plan and would correspond to the return of a certain number of similar conditions. The Mid-Atlantic Ridge would show a beginning of filling by folds trending across the Hercynian and Alpine plan and would initiate, by following the essentials of the Caledonian plan, a post-Alpine orogenic cycle. The long persistence of geosynclines, whose lifetime can encompass several cycles, with the preservation of the same general plan, is certainly an established fact. But the alternate return of plans that intersect each other on such a scale and for such long durations of time presents quite a different problem whose solutions is difficult in terms of fixist hypotheses unless randomness is assumed.

According to the mobilistic concept, the formation of the Proto-Atlantic becomes simply an old outline of the Atlantic that was generated by the traction and the thinning of a very old continent, both associated with drifts. The geosynclinal condition has been reached, as is well known; whether this has also been the case for the oceanic condition is not known, but fundamentally it does not matter; (304) nor is it important to know whether the geosyncline was simple, branching, or multiple. The Caledonian folding, a product of reverse drifts, of the nearing of systems that had been separated, reestablished the welding. It will probably never be known if it did reestablish it over the entire length: the present Atlantic conceals too many things. The folding of this great Caledonian branch is only the scar of an ancient wound, subsequently reopened with great consequences as shown by the gaping space of the present-day Atlantic. One of the advantages of mobilism in this matter is to reduce to the same concept the explanation of the former and the present Atlantic by using the intersection of the plans instead of being hindered by them.

It has been pointed out—not without reason—that the Caledonian geosyncline of Scotland and of Scandinavia behaved as a double chain with opposite overfoldings. On both sides, one sees along the margin clean-cut thrusts that in fact result from Caledonian basement folds. These folds have reworked the Precambrian of the very margin of the Hebridian and Baltic massifs, including the Cambrian or Cambro-Silurian coverings, and have satisfied their internal tensions by means of low-dipping great fractures and by going beyond any ordering. But the much more flexible style that prevails throughout most of the internal part of the Highlands, as well as in that of the Scandinavian chain, is properly the geosynclinal style of plastic behavior.

The *Mediterranean* and its chains present to the mobilistic theory a difficult field of application and a test that this theory must undergo if it pretends to have more than a passing acceptance. The smallness of the scale allows neither the large statistical approaches by which this theory is successful nor the easy unfolding of the greater deformations that it uses. The type of small-scale complication displayed by these structures, all this *three-dimensional puzzle* of which many pieces are deformed to the extent of being almost unrecognizable, and furthermore the very demanding stratigraphy, ready to lock up in its refined chronology each phase of some importance of the movements, require from the theory a calculated and cautious progress. Besides, it could not, without the help of a great number of operational artifices borrowed from the observed tectonics, penetrate very far into this problem, which is too close to the lower limit of the scales within its interest.

But, if one sees only the difficulties, no positive work will ever be accomplished. Therefore, let us admit, until further notice, that the history of the Tethys and of its Mediterranean segment has exhausted all the complications of the multiple behaviors that I have related to the geosynclinal condition, including the transitions to the oceanic and marginal conditions. Besides, I shall limit myself to a smaller number of sketches (Figures 13 to 27).

P. Termier[33] has shown in enlightening pages the Dinaric land overriding the Alps and passing under the Apennines.

A few years ago, I demonstrated that the ancient basement of the Dinaric land, as it occurs for instance in northern Italy, is but a fragment (305) of Indo-Africa. This meant that in the segment that includes the Alps, Africa overrode Europe or, in other words, the continent of Gondwana overrode Eurasia. I added that this overriding margin belonged to a large promontory that protruded to the north and complicated Africa and that to this promontory corresponded a reentrant of old Europe.

Indeed, if one proceeds from the crystalline basement of the Dinarides of Lombardy to the tabular lands of Tripolitania, following the plain of the Po and the Adriatic, one does not intersect or follow any object that would allow us to assume, for a certain past, a discontinuity of the continental block. What is intersected of the Dinaric chain in these areas is but a veneer of slightly shoved cover sediments. The embryonic ridges and furrows that stratigraphy allows us to reconstruct in the easternmost parts of the Dinaric arc, parts, moreover, that one does not cross but only follow, nowhere reach the importance of the axial zone of the Alpine geosyncline, or the Pennine zone. Therefore, they should be considered as symptoms of the flexible deformation of a slope that belonged also to Africa. The tabular fragments of Apulia, of southeastern Sicily, and of Malta outline the ancient continental junction, interrupted today by the abyssal areas of the Ionian Sea. The promontory of Gondwana, even if it ended in northern Italy, would be a very large feature indeed; but it does not stop there. The Austro-Alpine nappes, crystalline and sedimentary covers included, represent its most advanced and very highly deformed salient. These nappes were initiated by basement folds that developed their cylindrical ordering at the expense of the northern slope of Africa and perhaps, in part, of the basement itself—hence a rich differentiation in embryonic ridges and furrows, with some of the latter very deep at times. The offensive of these basement folds, sustained by the entire mass of Africa drifting to the north, became so powerful that the basement folds broke horizontally into huge rigid nappes and lost all ordering. In this process, they dragged along their unconformable cover, itself fragmented into spread-out, refolded, or wrinkled nappes; they overrode the true geosyncline, the Pennine zone, then the Helvetic zone—folded remnant of the cover of the European slope of the Tethys—finally, they dominated the very margin of the old Eurasia, visible in the Aar massif and reworked into basement folds.

This is the way in which, from the vicinity of Vienna to the Graubünden and further on in the upper Prealps, visible at the horizon of Berne, Neuchâtel, Lausanne, and Geneva, the superposition of Africa over Europe is shown.

The African promontory, in its more salient portion of that time, which shows today the deepest penetration, consisted of Hercynian folds, of a strip of African Altaids previously adjacent to the continent of Gondwana, of which it became an integrant part. The reentrant in which it is engaged and with which it displays affinities that may relate, in a complicated manner (306), to ancient driftings, is outlined by the large circumvallation with two loops that contains at present, from southeastern France to Rumania, the Alpino-Carpathic nappes. This filling of nappes does not allow us, outside the western Alps, to define the precise shape of the European reentrant, but it does allow us to estimate the dimensions and the general arrangement of the latter.

This penetration of Africa into the middle of Europe, of which I have spoken in the past under other circumstances, seems to me today—considering only concrete facts—to be unaccountable without continental drifts, particularly since the amplitude of the overthrusts, considering only what can be seen of it in the repeated soles of the Austro-Alpine nappes, is enormous even when we take into account some causes of error by excess that accompany this kind of estimation.

Here, another India has collided with Eurasia, but it was of smaller size than the Asiatic example; instead of going underneath Eurasia, it overrode the latter. Thus is defined, along the length of the continent, a *segment of central Europe* that strikingly resembles, with its Baltic massif, its Bohemian massif, and its fragment of Gondwana, the segment of central Asia with the Siberian, Serindian, and Indian massifs. But in Europe everything developed on a much smaller scale.

While drifting later on to the north, Europe tore away from Africa its promontory and preserved it between its folds. Indeed, this promontory was too much entangled with Europe to remain with Africa. Suddenly, the abyssal zone of the Ionian Sea and of the Great Sirte opened. Also suddenly, the Hellenic, southern Anatolian, and Tauric arcs, from Valona to the Gulf of Iskenderum, were torn away from Africa, and the large basin of the eastern Mediterranean appeared. Cyprus dragged a little astern. Again suddenly, the basin of the western Mediterranean opened up through tractions that profoundly disarticulated the preexisting chain. Naturally, although using the adverb "suddenly," I mean a certain length of time. The sial of the African promontory persists, stretched beneath the Adriatic, with a thinning that increases southward as disclosed by the increase of the positive gravity anomaly in that direction. The strange transverse furrow presented by the Adriatic on the 15th meridian is another expression of this stretching.

Besides, it is mechanically inconceivable that the major part of the Italo-Sicilian cordillera, or what was representing it, could have been at the time of the offensive of the African promontory as close as it is today to the Dinaric arc. This promontory, to have led such successful attacks over such a front, must have occupied an appreciably longer segment than the short distance that separates today the Italian arc from the Dinaric arc. It extended westward, massive and strong, at least as far as the 8th eastern meridian. Therefore, the major portion of the Italian precordillera was located farther west. (307)

What was the situation then of the present western Mediterranean? A compression was taking place, less intense perhaps in some areas than on the transverse alignments intercepted by the promontory but still of considerable magnitude. What objects were being compressed? The present-day Corso-Sardinian massif occupied, with a southwest direction—related to the present configuration—the long alveole of same dimensions which extends from the Gulf of Genoa to the Gulf of Valencia. Its western margin of today was directly adjacent to the present slope of the Riviera del Ponente, of France, and of a part of Spain. A portion of the granitic Corsica was facing the Esterel massif, and the south of Sardinia with its crystalline terrane, Cambrian and Silurian, was facing the old Catalonian massif of similar composition. Farther to the southeast—in the present configuration—the Balearic segment extended in a parallel fashion, the link between the Betic cordillera and the Alps. Still farther to the southeast, an Italo-Sicilian precordillera ran in a parallel fashion, and in a southwesterly direction; again farther away, almost aligned as today, a related structure of the Algerian Atlas and of the Saharan Atlas was outlined. The entire bundle consisted of elements parallel to each other on a large scale.

During the second phase, the one of the great tearings, the arrangement I have just reconstructed underwent profound injuries: cracks formed that cut into the cordilleras without much respect for their ordering; and buttonholelike depressions appeared, some to be enlarged into broad areas of thinned sial or of sima in which powerful eddies sucked eastward the Corso-Sardinian, Balearic, and Italo-Sicilian fragments. These fragments, although separated from each other, were still inserted by one of their extremities into the European continent.

The main center of suction was located south of the Alps and southwest of the Dinaric chain, in the sial of the promontory undergoing thinning and particularly in the underlying and adjacent simaʹ: it was a stern eddy with respect to Europe drifting northward. With time, the process of suction extended concentrically to the periphery, weakening with increasing distance. The deformations had, in the three dismembered and engulfed fragments, the character of an incomplete *release,* with rotation and change of direction of the free branches around the point of insertion.

The Italian branch rotated around its insertion north of the Gulf of Genoa; it swept all the azimuths between the southwest and the southeast passing by the south; it adopted, on the way, the great Siculo-Calabrian curvature; at the tail, the Sicilian extremity drifted 1,000 to 1,200 kilometers from west to east along the coasts of Algeria and Tunisia. Of the three sections engulfed by the eddy, the Italian fragment shows the greatest angular movement, the greatest displacement, the greatest deformation: they are all effects of proximity to the aspirating center. The Corso-Sardinian section, coming out from its alveole, shoved the extremity of the Balearic section, tucking it up to the southeast in the vicinity of Minorca. (308) The Balearic section rotated and oscillated a little around its insertion at the Cape of the Nao. The Corso-Sardinian section, continuing its rotation around its extremity inserted in the Gulf of Genoa, swept all the azimuths between the southwest and the south. Its movement was of the same style as that of the Italian fragment but less pronounced: hence an effect of the more upstream position in the eddy as a whole.

The Italian fragment, when confronting the promontory in the east, seems to have encountered there only an already strongly thinned out sial, into which it penetrated easily and which it incorporated, thus increasing its own mass. In this manner, the African promontory was reduced to dimensions smaller than those that during the immediately preceding span of time allowed it to play fully the offensive role that I have attributed to it. Therefore, the deformations of the Italian fragment in the thin sial take place at the earliest during the thinning; that is, while Europe, in the second phase, drifts northward while getting away from Africa.

The great offensive of the promontory, with the adjacent offensives to the west and east, corresponds certainly to the maximum of the Alpine paroxysm—to a certain portion of the Oligocene. The second phase corresponds to an immediately subsequent time span; the whole process took a relatively short time.

This great story in two phases is followed by the period of the replicas of the Alpine folding. The powerful tearings of the second phase have made any kind of large-scale synergy between Europe and Africa impossible. Numerous flexible behaviors are eliminated. But, others continue with moderation, and with renewals, in the style of the paroxysm but without ever being equal to its power. This qualitative renewal of the ancient style remains possible wherever by the occurrence of great but not too severely disjointed masses, flexible articulations generated by the paroxysm have subsisted. Thrusting processes continue along the external margins of many chains that increase by means of very young belts. Elsewhere, entirely new behaviors develop, generated by new general or local conditions. In summary, folding is far from having ceased, and nothing is more constantly flexible or more complicated in its behaviors than the chain of the Alps or that of the Carpathians during the Neogene and the Quaternary.

An important drift northward, a probable continuation of the one that was active during the second phase of the great story, animated Europe during the Neogene; at certain times, episodes of inversion may have occurred. I shall not deal with its folding effects. Among the traction effects is the trans-Aegean furrow, perhaps initiated—already near the end of the great story—by folding deformations. This furrow is filled by Oligocene and Early Neogene deposits. Distensional behaviors seem to have accentuated the almond-shaped configuration of the Pannonic depression and to have contributed to the establishment of the infratectonic complication of that region, where the areas of thinned sial, risen sima, and positive anomaly (309) predominate, entangled with a few negative zones. Other tractions, whose effects are clearly visible only after the Levantine Stage, produced the Aegean depression and the buttonholelike area of the Black Sea. Cracks, furrows, depressions, and "buttonholes" generated by these tractions display, as all kinds of folds do, *phase differences* between their different parts. A given degree of deformation reached along a particular transverse alignment will be attained only slightly later along another transverse alignment of the same object. It is with this sense of nuances that I have talked about the trans-Aegean furrow and the Black Sea. I shall assume these nuances for any folding or stretching behavior, for any deformation, but it is essential to express them clearly whenever dealing with lands like the surroundings of the Mediterranean, where the subdivision of geological time into short intervals is carried to a high degree of refinement.

The hypothesis of an ancient junction between Sicily and the Rif, near Melilla, presents stratigraphic and tectonic difficulties, which upon thorough examination have led me to abandon it. Probably, what is today the Rif followed the direction of the Betic cordillera, forming the southern margin of the Algarve. It assumed its present position, with the loop of Gibraltar, only after some movements pulled it away from its original location and tardily pinched it between Spain and Africa when they were getting closer to each other.

This segment of chain, separated from the Spanish Meseta and then tossed around by eddies, was gradually tucked up in plan and eventually collided with the Moroccan Meseta. The collision accentuated the shape of the loop and initiated the advance of the pre-Rif nappes, which took place, according to L. Gentil, M. Lugeon, and L. Joleaud,[34] at the end of the Helvetian. The entire deformation has certainly encompassed the first part of Neogene times; the Helvetian paroxysm has been its conclusion; and the closing of the South Rif strait, during the Tortonian, the late epilogue.

The Iberian peninsula, once liberated on the Atlantic side from its ancient connections to the northwest and west, most certainly dragged in the wake of Europe drifting to the north, with a certain amount of rotation to the southeast around the Pyrenean region and with a tendency to shorten the western Mediterranean—hence, certain complications of the eddies, which were active in that basin, and corresponding modifications of the shape of the released cordilleras. Temporary changes in the direction of drift might also, during certain epochs, have changed the pattern of these deformations.

My reconstruction of the puzzle of the western Mediterranean leads to that of the tectonics immediately preceding the great disjunctions.

At first, one sees that the Pyrenees, with their extension in Provence and in Cantabria, have merely relationships of vicinity and of intermediate synergy with the Tethys, their deformation being essentially internal with respect to old Eurasia, that is, to the continental block. The small embryonic furrows that occur (310) in this arrangement were deep in some places; they are miniature geosynclines, or basement synclines undergoing a particularly active filling, kept for a long time in a state of instability.

Granitic Corsica, the old Sardinian block, and the massif of Catalonia are the continuation of the first Alpine zone (Aar-Mercantour) and connect the latter with the southern margin of the Meseta, which is its homologue along the Guadalquivir and in Algarve. The lustrous schists of eastern Corsica belong to the Alps; this fragment is located along the connection—in the past almost rectilinear—between the Pennine zone of the Alps and the similar formations that are visible in the Betic cordillera. The upper nappe of the Apennines, with its comprehensive sequences (the so-called ophiolitic nappe), belongs only geographically to that chain. Tectonically, it is part of the Pennine zone of the Alps—with a strongly developped nappe structure in the original Alpine direction—subsequently affected by a large-scale backward folding that brought it to rest on top of the true Apennines, that of the Dinaric series, of the nappe of Sicily, and of its continuations in the peninsula. Indeed, these Apennines, as the Dinaric arc, consist of a strongly deformed segment of the African slope and of its sedimentary covers. The backward folding of the ophiolitic nappe is but the exaggeration of the one that is visible along the more northern transverse alignments at the extreme internal margin of the Pennine Alps.

A lot could be said about the embryonic movements, about the deformations of the Mediterranean and of its chains, or about the art of reconstructing, for each epoch, the strongly movable framework—continuously deformed and without any fixed point—of stratigraphy and of paleogeography. I deliberatly omit all this. I shall say only that the large African promontory often has acted in the same direction as it did during the great Oligocene offensive, even with interruptions of fighting but with a certain insistence upon playing the same role, thus generating within its mass and around it a style that forecasts that of the great story. In the Upper Jurassic and in the Lower Cretaceous, numerous Andean movements originate from such conditions; often they linger until the eve of the Cenomanian, as in the eastern Alps, which belong to the promontory itself, and in the Getic nappes, where the deformation was strongly developed. I shall only mention here the pre-Lutetian movements whose complex relationships with the great Laramide folding has been previously mentioned.

The independence displayed by the marginal thrusts of the Alpides, with respect to the ante-Alpine folds of the old Europe, is well known. With respect to the Alpine basement folds, this independence is but slightly less important, a remarkable fact. Trending southeast, the group of basement folds that includes Bohemia and the Sudeten goes under the Carpathians in the same way as the old folds do. More generally, the direction of the basement folds from the second to the fifth zone is oblique with respect to that of the Alpides. The strong (311) sial of the old continent has been folded according to its own laws, as a function of the different driftings it underwent and without paying attention to the more delicate constitution of the depths of the Tethys, with their thinned sial, their new deposits, their sima, and their eddies regulated by different laws. Quite often, basement folds while extending themselves have reached at right angles or obliquely the slope of the continent, but none of them has been able to continue appreciably within the geosyncline. Basement folding, independent of the geosynclinal deformation, took place before, during, and after the paroxysm.

The *Mid-Atlantic Ridge,* similar to a wreck of sial tossed around and deformed in the sima, seems to be the heritage of the time in which the New World was beginning to separate from the Old. It seems to consist, as previously said, of the remains of the sial stretched at the bottom of the S-shaped furrow along which the disjunction was being prepared; of masses of sial collapsed from both vertical margins, with sliding or dragging of some terrigenous deposits, even of melted sial.

But there is more to it, and it is appropriate to stress, on the one hand, the role of certain connections that the ridge has kept with the Old World and, on the other hand, the presence within the ridge itself or along its margins of conformations that can, most certainly, be related to the folded zones that trended from the Old World to the New.

With the passing of time, the two Americas, drifting westward, left behind them the stretched band that became the ridge. This ridge was, however, to a great extent disjointed from the Old World and it is floating today in the center of the Atlantic sima. Nevertheless, this latter disjunction is not complete: the Walvis Ridge, a kind of oblique dike floating on the sima, still connects the Mid-Atlantic Ridge with South Africa. The junction, which is suggested by a few rare soundings only, is demonstrated by the thermal stratification of the ocean. Such oblique bridges, also of distensional origin, often accompany the crevasses of glaciers. Less certain, in the present state of the soundings, but still possible, is the continuation of the ridge into Iceland—hence the probable junction of this ridge with Europe by means of the Wyville-Thompson Ridge. These connections, particularly the first one, which is unquestionable, are sufficient to make very unlikely another hypothesis about the origin of the Mid-Atlantic Ridge, namely, that this object is the remains of an ancient sub-Andean lens completely released by the two Americas and which would have risen, astern, in the Atlantic sima. Besides, such a hypothesis is not in agreement with the existence of some sandy deposits that have been dredged at several places of the submerged ridge.

While undergoing distension, the band that was to become the ridge inflicted the same treatment to the segments of folds and of chains crossing that band from east to west, and many of these segments must have passed there connecting the present Old World to the New. These folded objects, stretched lengthwise (312), have some chance to be preserved—although more or less deformed in plan by eddies passively undergone—in the shape of inclusions or appendices, transverse, subtransverse, or twisted, inside the major part of the Mid-Atlantic Ridge or along its margins as fringes. Such distended fringes occur in great number, as east-west trending ridges and furrows, over the large area enclosed between parallels 47 and 54 and meridians 20 and 40, according to the bathymetric contours given by Sir John Murray.[35] One can see in these broad conformations, aligned over more than 2,200 kilometers along strike, the elongated, stretched, thinned out, and consequently depressed remains of the basement folds that extended from Cornwall and the Armorican massif to Newfoundland, the Taconic chain, the Appalachians, and the Piedmont, across the segment today distended, collapsed, and abandoned.

The progress in sounding by the new acoustical method is of so much better resolution than the ancient techniques that it will certainly reveal with more precision the configuration of these objects. It will allow us to discover, in other latitudes, new fringes adjacent to the Mid-Atlantic Ridge, in part to be attributed to the transatlantic connection of other bundles of folds. In a more general manner, it will have a decisive influence on the location and the solution of the problems pertaining to the submerged tectonics. Meanwhile, one can state that if it is useless to pretend to disclose the occurrence of narrow folds in the great depths, by means of the ancient methods of sounding, this is not the case for the stretched remains of basement folds in the better explored portions of the ocean. The great width of these objects does not allow them to escape interpretation as soon as the network of soundings reaches a certain density. Thus, the concept of basement folding introduces, in the problem of the transatlantic connections between folded bundles, an element accessible to bathymetric control, which is not the case for ordinary folds.

Similarly, one may also be allowed to interpret the twisted wrinkles of the Azores, enclosed within the Mid-Atlantic Ridge, as the stretched and distorted remains of new chains that were rising from the Mediterranean through the interval between the Spanish Meseta and the Moroccan Meseta. One may think immediately of the continuation of the Betic and Rif cordillera, today looped in the opposite direction, and of the Italo-Sicilian cordillera, presently drawn to the east, as previously mentioned. In the interval between the Mid-Atlantic Ridge and the Old World, other wrecks seem to build—as long as they are not volcanic growths—the group of shoals that are aligned or scattered from Madeira to the Iberian peninsula, the Seine seamount, the Dacia seamount, and the substratum of the Canary Islands. The part of all this (313) that pertains, on the one side, to debris of Mediterranean cordilleras that were dragged westward by the Mid-Atlantic Ridge and then released and abandoned to their fate or, on the other hand, to any fragments released by the main part of the ridge cannot be established today. But the first interpretation is probable for objects of elongated shape.

It is premature to eliminate, as has been proposed, the hypothesis of a connection between the Alpine chains of the Mediterranean and those of the region of the Antilles. Even if it is true that the arc of the Lesser Antilles outlines a junction between the great folded bundles of the west of both Americas, nevertheless, the zones that form the island of Trinidad escape straight eastward, in the direction of the Mediterranean. However, the occurrence at Trinidad and in the cordilleras of Venezuela of a strong frame of Andean age and of a properly Alpine frame, superposed on the former by means of a strong unconformity, introduces—with respect to the Mediterranean chains, with their much smaller Andean deformations—a difference that it would be imprudent to underestimate.

There is no region of the *continent of Gondwana* that did not move by basement folds during the Alpine cycle. One should distinguish the marginal basement folds that encroached upon the margins of Gondwana, while keeping a relatively narrow style in which the average radii of curvature predominate, and the internal basement folding, either positive or negative, of very broad style, of large radius of curvature, and of such general extent that all the remaining portions of the old continent have been involved in it.

The marginal basement folds of Gondwana include the Himalayan zone, the Australian Alps, and the major part of the Antarctic Andes and of the South American Andes.

The basement fold of the Himalayan zone was definitively initiated in the old sial of Gondwana, by its collision with the old sial of Angara; both influenced each other, first in an indirect manner across the sial of the Tethys reworked into marine deposits, then in a direct manner by the substratum of the Tibetan intumescence.

The basement folding of the Australian Alps, which reworked old Caledonian and Hercynian folds, seems to have been generated mainly by the indirect resistance of the sima operating across folded wrinkles that presently belong to the Australasian chains released to the east and northeast—New Zealand and others—but that at that time were adjacent to Australia as marginal chains enclosing basement folds.

The basement folds of the South American Andes are due essentially to the resistance that the Pacific sima presented to the westward drift of Gondwana, and later on to that of South America, when it became separated from Gondwana by the opening of the Atlantic. One cannot say as yet, as was done for North America, how many main generations of basement folds have contributed to the construction of the South American Andes during (314) the Alpine cycle; the phases have, at least, been numerous. Taking into account the reused dead material, one can distinguish Alpine basement folds consisting of Paleozoic folded material, which build the eastern chains of Argentina, Bolivia, and Peru, and Alpine basement folds that build, along a more western and longer zone extending from Venezuela to Cape Horn, the major part of the Andes proper: their dead material consists predominantly of Andean folds, that is, folds that were new at the end of the Jurassic or in the Early Cretaceous. The new tonnage, younger than the Andean folding, appears in the cover of these Andean-Alpine basement folds. In the present state of knowledge, one cannot say whether this tonnage is concentrated, in some places, in entirely new coastal chains. But it is certain that the whole of the new tonnage, in South America as well as in North America, weighs little in comparison with the reactivated tonnage.

With respect to the cordilleras of Antarctica, the little that is known of them at present prevents any discussion except in terms of a general analogy with the South American Andes.

The basement folding of the interior of Gondwana reaches, in total, the most impressive tonnage. The latter encompasses, except for the cover folds, all South America outside the Andes, Africa less the Atlas chains, Arabia less the arc of Mascate, peninsular India, Madagascar, the old Antarctica, and Australia less the chains of the east.

The distinction between positive and negative basement folding and that between effects of compression and of distension being rather delicate in certain areas, it is appropriate to make in that respect some remarks that are, besides, of general implication.

The great intumescences can be attributed to anticlinal basement folding, by compression, with increasing ease as they are better ordered and as their whole displays a better expressed unity of plan. Whenever such objects have undergone sufficiently clear disjunctions, with the preservation of almost congruent segments, the interpretation will thus be rather facilitated than hindered. But, if the ordering is deficient or if a chaotic conformation appears due to deformations maintained at the incipient stage, then the above-mentioned distinction will become difficult on the basis of the known facts. It will be the same whenever one can assume that the ordering and the unity of plan were destroyed subsequently, either by means of widely distributed and great distensions or by secondary isostatic behaviors.

The same prudence should be used, with a few additional precautions, with respect to the interpretation of large basement synclines that can be simulated, to a certain extent, by depressed zones due to tractions. But when depressions and intumescences belong to the same unity of plan and when this unity, furthermore, displays with evidence the mark of a compression, of a folding, any confusion can be eliminated. This will happen, for instance, (315) when the above-mentioned unity is related to a large-scale virgation because nobody would think of attributing such a general arrangement to distensional behaviors.

Furthermore, the idea of a basement syncline does not necessarily imply, in my mind, that of a downward flexure. This flexure may occur as a secondary feature, under the influence of the deposits that are called in, so to speak, by any depression surrounded by adjacent highs. I am inclined to assume that this flexure may occur originally, for small semi-rigid objects, by being restricted to their upper parts and by decreasing in importance downwards. But, for great objects, the downward flexure would require such a great work performed against the principle of Archimedes that it is foolhardy to think about it except in the case of secondary overburden.

In general, the behavior of the great basement synclines, with respect to the vertical effect, seems to me to be expressed by a rising, but a rising of less magnitude than for the anticlines. This may imply, beneath both basement synclines and anticlines, a downward increase of the lenses, an increase of the thickness of the continental sial, but of lesser amount in the case of synclines. One can visualize at least two cases in which depressions of synclinal or tabular shape, skirted or framed by basement anticlines, display little or no original vertical effect. Both cases assume a compressional effort that is satisfied by an increase of thickness essentially restricted to the anticlinal regions, for both the visible parts and the infratectonics. The first case, conceivable only for narrow depressions, occurs when the bottom of the depression behaves almost as if it were a simple transmission organ, while the anticlinal wrinkles grow upward and downward. The second case, which for a certain period of time appears to be that of extensive regions—exposed or tabular—results from the fact that the operating basement fold, which temporarily concentrated in the anticlinal regions, has not encroached, or has momentarily ceased to encroach, upon the major part of these intumescences. Besides, I think that these are boundary cases that are worthwhile considering as an analytical approximation, but it is not possible to say at present whether they can develop with all their conceptual precision. I shall not discuss the cases in which sub-anticlinal lenses extravase partially toward the deep-seated zones of adjacent depressions nor those in which the deformations occur within a previously thinned sial. Furthermore, I shall also leave aside, even more deliberately, the secondary behaviors of all kinds.

These precautions taken, I shall now turn again to concrete objects.

Most of Africa displays a grandiose unity of plan, that of a great and powerful virgation of basement folds, which is of the first type and additionally encompasses Arabia and the extra-Andean part of South America. The visible intumescences, in the heart of which are exposed crystalline rocks; (316) the warpings of the sedimentary covers deposited during the Alpine cycle; and, finally, the curved peneplains allow us to sketch, in a first approximation, within the limits of present knowledge, the major features of this *internal virgation of Gondwana* (Figure 6).

A first virgation branch, mainly anticlinal and complicated during its late development by more than one distensional behavior, encompasses South Africa, most of eastern Africa, Ethiopia, and the great intumescence of both shores of the Red Sea, in Sudan, in Egypt, and in Arabia.

A second virgation branch, mainly anticlinal and consisting of intumescences with a sometimes sporadic distribution, although sufficiently aligned on a large scale is visible in the Darfur, in the Tibesti, and in the masses that form the so-called central Saharan massif, where the unfinished growth of this branch to the west is shown by somewhat chaotic and confused features. Following the second branch in the opposite direction, to the southeast and south, one sees it moving closer to the first one, with which it appears to display close relationships starting at the 5th parallel north. Farther south, the great general intumescence of eastern Africa seems to contain, along its western margin, some elements related to the second branch, while most of it belongs to the first. Therefore, this intumescence may be in a sense considered the common trunk of the first and the second branch.

Welded to the first branch and to that trunk in the extreme south, a third virgation branch, mainly anticlinal and affected by the Atlantic disjunctions, follows the coast of the ocean across the lower part of the basins of the Orange, the Congo, and the Niger, continuing after a general inflection to the west, north of the Gulf of Guinea, as far as Sierra Leone and the vicinity of Conakry. Farther away, this branch continues in the massif of Guiana, which is its free extremity, disjointed by the Atlantic fractures and carried westward by the drifting of South America.

A fourth protruding mass follows the east coast of South America, from the estuary of the Rio de la Plata to the vicinity of Cabo de São Roque; it reaches Ceará in the north. The fourth branch displays close affinities of behavior with the third branch; it may have belonged to it, before the disjunction, without there having been any great intermediate basement syncline.

A fifth group of intumescences appears in the very center of South America. Although it is impossible at present to establish precisely its shape, the few known features disclose alignments to the north and northwest that indicate with respect to the previous mass a divergence and an opening in the same general direction as of the other branches of the virgation.

In Africa, the diverging branches open toward the north, northwest, and west, and leave exposed, between the first and the second branch, the large space gradually increasing to the north (317) that corresponds, as a whole, to the Nile basin and to the Libyan desert. The second and third branches, while separating from each other, similarly open the depressed space shown by the Chari and Chad basin, and after a change of direction of the plan to the west, shown by most of western Sudan. Following in the opposite direction this chain of homologous tectonic depressions, always located in front of the third branch of the virgation, one finds first, beyond the Chari-Ubangi threshold, which expresses an axial culmination, the great basin of the Congo, with its filling of sandstones, then another axial culmination located in the vicinity of the 12th parallel south, which leads to a last depression, stretched out between the 12th and the 23rd parallel and occupied from north to south by the upper tributaries of the Zambezi, the internal basin of the Ngami, and the Kalahari. Farther away occurs the junction of the third branch with the more internal elements of the virgation, which an uplifted basement syncline—the plateau of the Karroo, of the Orange, and of the southern Transvaal—separates into two bundles of predominantly anticlinal behavior. These large African basins constitute, in one or the other indicated direction or in more than one of these directions if behaviors on a smaller scale are considered, very broad basement synclines or protected tabular areas moderately warped, which in principle amounts essentially to the same as pointed out earlier. The horizontality of the covers, which prevails over large areas, does not contradict this concept; it indicates a powerful deformation with its characteristically quiet spaciousness.

Therefore, most of Africa belongs to a virgation of the first type whose central segment covers approximately the space enclosed between the 5th parallel north and the 25th south; what is farther south seems an incipient right wing; what is farther north is the main part of a broadly spread left wing, of which certain branches extend into South America. The convexity of the system, according to its main features, faces the east, and the wings lag behind, tardily: hence the concept of a compressional effort that is predominant, unilateral, more or less continuous, or renewed, more or less helped or antagonized, according to times and places, by reverse or distensional stresses, but directed in the center to the east, at the extreme right to the southeast and south, on the left wing to the northeast and north—in summary, always from the inside to the convex outside of the sheaves.

The shape of the branches of the virgation, when South America is being brought back against Africa, in the position before drift, displays a general conformity with the South American Andes, a sure sign of a folding synergy that involved both stories before the great Atlantic disjunction. The curves of the African branches belong obviously to the same family as the curvature of the Andes on both sides of the Gulf of Arica; this conformity goes very far for the third branch after its change of direction in the Gulf of Guinea; it is well expressed in the relationships of the second branch with the main portion of the virgation; and it is still visible but more faded away in the general curvature of the first branch. Therefore, a plastic push (318) originating from the west, to which the entire frame of Gondwana reacted, has been felt across the continental mass, and its effects on the curvature of the plan slowly decreased eastward. In order to account for a deformation of this size, one must assume, as for any kind of basement folding, the internal deformation of the sial and the friction of the basal sima. To these factors was added, before the Atlantic disjunction, the resistance of the Pacific sima in front of Gondwana drifting to the west, that is, at the bow of what is now South America. This factor explains not only the general conformity between the Andes and the virgation but also the unity of plan that appears in the latter. In vain one would attempt to explain all these homologies without synergy between the Andes and the virgation. These interdependent deformations consisted of many phases, and the occurrence in the north of Tanganyika of Andean movements, demonstrated by the unconformity of the Middle Cretaceous over Jurassic formations, indicates that this synergy, far from being an illusion, has encompassed at least all the width of South America and of Africa still welded together.

With respect to the imagery of the movements and taking into account the differences between basement folds and ordinary folds, we can see that the relationships between the elements of virgation located south of the Amazon and the Guiana segment of the third branch are similar to those between the Afghan virgation and the Persian virgation but less strongly expressed. The elongation of these elements toward the north and northwest, more or less perpendicular to the Guiana branch, appears as if incomplete, and the broad basement syncline of the Amazon, which extends parallel to this branch, indicates such a situation. From the same viewpoint, the relationships between the third branch and the Colombian and Venezuelan Andes are similar to those between the Persian virgation and the northern Iranian arc, if details are neglected. The elongation of the Guiana branch to the west also is incomplete.

The pre-Andean depression, a long and broad basement synclinal zone, is first expressed by a part of the Orinoco valley; an alveole trends through Patagonia and occurs again, in a state of preservation or of disjunction, which the ice cover does not allow us to define, between old Antarctica and the Antarctic cordilleras, from the inner part of the Weddell Sea to that of the Ross barrier, then in Australia from the Murray basin to the Gulf of Carpentaria. The Australian Alps, considered as basement folds, correspond to a part of the South American precordilleras with dead Paleozoic material. The more external Australasian chains, with their Andean deformations and their properly Alpine replays—involving reactivated tonnage and new tonnage—correspond to the true South American Andes with their reactivations. The pre-Andean depression, the marginal basement folds of Gondwana, and the new chains thus build a triple ring of variable width around the western, southern, and eastern margins of the continent; these orientations are, as always, given with respect to present locations. (319)

The degree of homogeneity of the major part of Gondwana is much greater than that of old Eurasia; it explains the ease with which, before the Atlantic disjunctions, the great internal virgation developed according to the laws of the first type, imposing them upon the major part of the continent.

After completion of the South Atlantic disjunction, the deformations of the virgation may have been suppressed or deeply changed very close to the newly cut-out fronts. However, it is reasonable to assume that in the interior of South America and of Africa, they have been reactivated by more than one phase, while the old articulations allowed the replay of the ancient movements in an unchanged or slightly modified style. Indeed, it is natural to think that fragments of such size remain capable of undergoing basement folding even after their disjunction. Thus is explained, in particular for eastern Africa, the youth of the last additions of a part of the intumescences.

None of these folding behaviors prevents the consideration—before, during, and after the major part of the principal disjunctions—of as many distensional effects as are required to explain the observed facts on much smaller scales. For instance, there is nothing contrary to the hypothesis according to which a portion of the African rift valleys or the Red Sea with the Jordanian graben and its Syrian continuation could have been generated by tractions generally oriented along the parallels. In order to make everything agree, it is sufficient to let space and also time act with the successions and alternations of behavior that it allows to operate.

Only the most general features of the axial behaviors proper to the intra-Gondwana virgation can be outlined in the present state of exploration. A transverse depression occupied by Karroo sandstones overlain by basalts is displayed in the third branch along the African coast and in the interior, between the 19th and the 21st parallel south; it occurs also in southern Brazil with a similar structure and in a congruent position, between the 28th and about the 30th parallel south. This arrangement clearly expresses the synergy, the intimate mutual dependence of the folding stresses, which affected this segment before the disjunction. The basin extending from the Kalahari to the upper Zambezi and that of the Congo are simultaneously transverse and longitudinal synclines if one neglects details; both are structural umbilical areas separated by a more or less transverse zone of culmination corresponding mainly to the 12th parallel. Farther north, a new zone of culminations includes the Ubangi-Chari threshold and most of the Uadai in a northeasterly direction. A major transverse depression partially occupied by Cretaceous and Nummulitic deposits extends from the Gulf of Guinea in a northeasterly direction through the Benue basin into Chad, beyond which the same behavior is emphasized by the tabular region of Ennedi between the tectonic highs of the Tibesti and those of the Uadai and the Darfur. The Chad, located at the point of intersection of this transverse depression (320) and of the longitudinal depression that separates the second branch of the virgation from the third, seems to correspond to a new umbilical area. Another transverse depression is represented by the region of the lower Niger; the Nummulitic sea penetrated into it. Once passed this entrance, it spread out in a broad fashion north of the third branch over a part of Sudan. The role played by the axial behaviors of basement folds with respect to the opportunities provided to the fluctuations of transgressions and regressions was therefore not inferior to the role played by the depressions of distensional origin through which the disjunctions began. Traces of these depressions have been preserved in the narrow coastal plains with marine sediments, often dislocated by numerous steplike fractures, that fringe most of the old blocks that originated from the breaking up of Gondwana.

I shall not discuss the morphological and stratigraphical consequences of this huge basement folding as much as one could do it today. But, it is fairly obvious that most of the course of the Nile is inherited from a consequent drainage over a basement syncline. Among the subsequent changes, which are naturally numerous, some are simply epigenetic in origin, while others are related to the continuation or to the replay of tectonic stresses. Thus, while broadening westward, the Eritrean intumescence, or the first branch of the virgation under discussion, has caught up with the Nile at different points along its course and forced the river into several antecedent behaviors that make it encroach on crystalline rocks. The general organization of the drainage in the Congo basin displays a close relationship with the structural umbilicus shown by the basement folds of that region; this is also true, with respect to the more southern umbilicus, of the general arrangement of the drainage in the basins of the upper Zambezi and of the Ngami. The topographic crest between the two groups of basins corresponds as a whole to the structural crest marked by the axial culmination of the 12th parallel. The lower courses of the Congo and of the Niger seem to have kept their respective positions by means of rejuvenations that the rising of the basement fold of the third branch generated without being able to keep under control: hence the encroachment on the older rocks in several places. Therefore, the pre-Andean depression is emphasized, over long distances, by a portion of the course of the Orinoco and by the arrangement of most of the tributaries, with meridian direction, of the Rio de la Plata. It is superfluous to add, in such a cursory examination, that the dependence of the major hydrographic features on the original structural conditions is in most cases indirect because too many episodes were intercalated and too many obliterations occurred.

Thus, the internal deformation of Gondwana consists essentially of a virgation of the first type that is the largest known virgation on the Earth.

The deformation of a great mass of sial results from a synergy of plastic forces and not from the transmission of efforts within a rigid medium. Actually, the heterogeneity of the upper parts determines most of the folded configurations. A small-scale mosaic, complicated and with numerous sutures between media of different average plasticity, like Eurasia, does not react in the manner of a less heterogeneous continent, such as Gondwana. In the former case, the distribution of the effects—the foremost being basement folding—is of a small scale that is controlled by that (321) of the mosaic itself, as if traced over it. Hence the almost always compressed style of basement folding in Eurasia, its moderate radii of curvature, its excellent ordering, its close dependence on the sutures, and the predominance, as in the latter, of the east-west direction. Hence, also, the small-scale distribution in Eurasia of the Andean and other movements developed in synergy with the great events that so deeply jostled the bow of the continent, namely, the Pacific margin of North America. The Andean effects, and more generally the effects originally generated or initiated in synergy with the bow stresses, have perhaps not been more important in Gondwana than in Eurasia, but they are more visible and more widely distributed because Gondwana, a relatively homogeneous continent, led not to such an advanced subdivision. Thus are explained the strong unity of the intra-Gondwana virgation, its dimensions, its general meridian trend, and the visible influence it has far away in the rear parts.

I shall leave aside the basement fold, mainly of anticlinal character, with a large radius of curvature and with fractures, that must have affected the major part of central and western Australia, the old Antarctica, Madagascar, and the Indian massif before the great distensions that separated these fragments from the rest of Gondwana and that generated or accentuated, by stretching or tension fractures, the depressed condition that these objects display over great lengths along their present margin. These great disjunctions, which are stern effects with respect to Gondwana drifting to the west, have, so to speak, peeled off twice the rear part of the continent; the first slice includes the Australo-Antarctic block; the second, more internal, corresponds to the Indian-Madagascar zone.

The so-called geosynclinal depression of Mozambique, stretching from the channel bearing that name to the sea of Oman, is essentially of distensional origin, both with respect to its first observable deformations, which are of Liassic age, between Africa and Madagascar, and to its last important ones. I have pointed out a few Andean deformations, probably of folding nature, along the margin of eastern Africa. Many alternations of compression and distension may have affected this area throughout the Alpine cycle, but it seems that no new folding of very large tonnage has taken place in that geosyncline; the rather quiet condition of the sedimentary covers in Africa and in Madagascar, as well as along the western margin of peninsular India, a quietness interrupted only by distensional fractures, demonstrates this situation.

XXX. Glances at East Asia

We can certainly give credit to A. Wegener for having noted that the arrangement in repeated festoons, so obvious along the margin of the continental mass of East Asia, is a type of very large folding in plan resulting from an effort (322) oriented parallel to that façade of Asia, which has thus shortened from about 11,000 to 9,000 kilometers. Besides, it is probable that something survives beneath this arrangement from an ordinary segmentation older than the shortening. As to the cause of this situation, one should assume, in my opinion, the lateral repercussion of the great Indo-Angaran frontal compression. Indeed, the largest of these protuberances is the nearest to India: it is the one that includes Indochina and western Indonesia. Even if we do not restore to it the released festoons of the Philippines and of eastern Indonesia because of the doubts we may have about the time of their release, it remains the largest of all. At present, it is impossible to exclude the hypothesis of drifts parallel to the eastern façade of the continent, drifts that would have temporarily participated in the production of this large-scale festooning.

The shortening of 2,000 kilometers undergone by East Asia leads us to believe that the algebraic sum of the drifts of India and of Angara Land was of that order of magnitude. Other inferences drawn by A. Wegener from other considerations raise this figure to about 3,000 kilometers for the drift of India toward the northeast only. The different procedures of estimation, including the one I have indicated, contain implicit postulates that lack of time prevents my analyzing here and that one cannot at present accept or reject with certitude. Nevertheless, the agreement remains valid as to the order of magnitude. But, the discussion of the unraveling of the folds of the Indo-Serindian space cannot lead at present to a reasonable estimation, and I have explained why.

I shall now return to East Asia.

There is no reason for the time being not to admit the retractions of the continent that generated the stern stresses: the opening of marginal seas, releases of island festoons, back eddies in the deep furrows at the margin of the ocean. The island festoons display many features of a more ancient order, which is that of bow stresses, of foldings, an order that one cannot say has been completely abolished. Naturally, I do not speak of the folds that can be seen everywhere; I mean the large-scale arrangement of the arcs with their segmentation, their virgations, and their linkings. One cannot attribute all this planimetry to the deformations in plan produced by stern stresses; it is essentially the survival of the ancient plan in the new one. This kind of survival with deformation is one of the fundamental features of island festoons. The change to stern stresses seems in most of the cases to have increased the curvature of the arcs, as it has increased, compelling them to open, the angles of linking and of virgation.

The virgation of the Alaskides, located at the extreme limit of the Asiatic plan and much less affected by the stern stresses than that of the Philippines, (323) gives us the approximate picture of a previous state of the latter. With the exception of the Aleutians, west of the 166th meridian west, all the arcs that build the virgation of the Alaskides have remained connected; they become gradually pressed together to the east, under very acute angles. There, the left wing is located. The virgation is of the second type, with a central transverse alignment corresponding to the 167th meridian west, the Aleutians having been brought back to the internal slope of the Bering Sea. The Philippines show, with a visible tonnage that is essentially new, a strongly developed disjunction of the virgation, which is also of the second type; the very abrupt convergence of the branches, when nearing Luzon, cannot be attributed entirely to the initial bow stresses. Compressing the whole structure, one reproduces very closely the plan of the virgation of the Alaskides, with its acute angles and its narrow arrangement.

These two virgations were initially generated by the oblique resistance of the sima at the two left wings, as bow stresses. The original relationships between the Kuril Islands and Hokkaido and between the Ryukyu Islands and Kyushu seem to correspond to a very acute echelon pattern between branches of virgations of the second type rather than to proper linkings. Stern stresses have increased the opening of the angles.

The fixist imagery can certainly account for these peculiarities by playing with segmentary variations of the power and of the resistance of the Pacific bottom. But fixist theories, as pointed out above, are not at all in agreement with what is revealed, on a higher scale, by the more general fluctuations of the Circumpacific tonnage.

Southwestern Japan, with its narrow Mesozoic bands often folded toward the Pacific and pinched between large basement folds or shoved by them, shows more reactivated tonnage than new tonnage.

On the shore of the Sea of Japan, the great island of Honshu displays Tertiary deposits often affected by backward foldings, with overturning to the northwest or west. The apparent inversion of the direction of thrust can be explained, without changing the general direction of the bow stresses, by assuming a downward displacement of the maximum effort in time. One can also assume stern stresses.

In general, it is credible that stern stresses, while compelling a festoon to accentuate the curvature of its plan, can develop in it all kinds of folds, particularly, true transverse folds, on the internal side, which undergoes compression; whereas traction effects appear on the external side.

The great basement folds of northern Honshu and of Hokkaido— Abukuma, Kitakami, Hidaka—show an oblique echelon pattern with important longitudinal offsets. This arrangement, which is displayed also by folded elements of variable nature in Sakhalin, seems to result from a longitudinal traction that the continent (324) imposes from the north upon the festoon in accordance with a pattern of stern stresses. The fixist imagery can find in it the incipient stage of a virgation of the second type, with an echelon pattern, but cannot progress any further. The mobilistic imagery can consider both viewpoints as true, assuming a succession or an alternation of bow and stern stresses.

In front of us, the Pacific Ocean stretches beyond sight. The few clear concepts that have been derived from it, by playing with statistical averages and the law of large numbers, should not make us forget the significance of the numerous mysteries that still remain concealed under its deep waters.

XXXI. Conclusion: Mobilism and Geologic Reality

The idea of *framed folding*, whose first expression is fairly old and which has become in the interpretations of Suess one of the refined aspects of his work and one of the highest achievements of classical tectonics, leads me to my last hypothesis.

I would like to show that the idea of framed folding is capable of such a general applicability that the mobilist theory itself, through its fundamental postulates and in spite of a few subversive appearances, is really a particular case of the former idea.

The framing matter, through its degrees of consistency, and the shape of the frames, through its distribution in space, regulate the degrees of freedom or of constraint that occur in any folding. Interacting with the tangential stress and with the degrees of consistency of the framed matter, these same factors determine the entire behavior of the traction effects and of folds of any kind, in volume and in time.

Such is, concentrated in a few words, the phrasing of this generalization. Now, here are the steps leading to it.

On the lower scale is cover folding, whose frame consists of the higher points of the immediate vicinity: basement folds, voussoirs, and even new chains.

On the middle scale are the foldings of the geosynclines, whose frame is outlined by the continental jaws with their slopes and, farther away, by the basement with its basement folds, often reworked into voussoirs.

Conclusion: Mobilism and Geologic Reality

On the upper scale, basement folding operates, involving the entire continental mass and whose frame is, if the mobilist theory is correct, nothing but the sima.[36] (325)

With each enlargement of scale, these expressions of framed folding increase in extent and in depth and, consequently, in tonnage. Cover folding involves only a thin film of sedimentary or volcanic rocks and at the best would not be able to extend beneath the great basal unconformity. Geosynclinal folding has by definition a smaller *penetration* than that of basement folding since it usually begins within a thinned sial. Basement folding—including under this term all the deformation of the sial that leads to the production of visible basement folds—has the greatest possible penetration since it operates within the entire thickness of the sial, which is, with numerous and important variations, of the order of 120 kilometers.

Therefore, the known types of foldings are arranged in space into encased frames to which it would be vain, however, to attribute too precise boundaries everywhere.

The hypotheses peculiar to mobilism that make this concept in its essence one of the types of framed folding belong to the first fundamental assumption of that theory and pertain to the nature of the medium that frames the sial undergoing folding. These hypotheses imply the concept of a frame of sima that yields and allows the greatest displacements of the framed object or resists as an ordinary frame, generating the folding of the sial, or, most often perhaps, allows delicate compromises between the two behaviors.

The original mobilism, in spite of the few and rather limited destructions it brings, is therefore neither the violent revolution nor the intellectual catastrophe that a few persons seem to fear. It is highly positive and constructive. One shall never know how the great founders such as Elie de Beaumont, Charles Lyell, James Hall, Edward Suess, and Marcel Bertrand—these minds naturally led to tectonics, which I cannot think about without feeling deep gratitude toward the past—would have received it. But I do know that Marcel Bertrand in his last bold schemes came close to it and perhaps even guessed it. Most certainly, these great founders would have explored mobilism in all its aspects and outlined its relationship with the tectonics they were creating.

The fine nuances visible in the upper parts of the old sial, subdivided into broad zones of unequal average plasticity, are expressed by reactivations that also demonstrate the concept of framed folding, of an order (326) of magnitude that connects the middle and the upper scale.

The generation of deep-seated lenses tends to equalize the average penetration of geosynclinal chains and that of basement anticlines. Besides, basement folding can affect a thinned sial, but this is only a particular case. Marginal chains, like all basement folds, have the sima as their frame, overlain or not, according to locations, by the remains of a thinned sial. Their scale is the same as that of geosynclinal chains.

If the original mobilism can easily be considered, following a simple reasoning, as a particular case of the concept of framed folding, this should also be the case, and the more rightly so, of that type of mobilism that I have presented loaded with all the aspects of concrete tectonics. While undertaking this delicate operation, I have rejected, after examination, all the suggestions of the vague eclecticism that would have attempted to reconcile irreconcilable terms in the twilight of powerless combinations or in the seesaw games of a delightful skepticism. These little artifices have never led to anything serious, regardless of the situation considered. I have simply attempted to draw in full light the map of the compatible, and it turns out that such a field was still very extensive.

With the original mobilism, a new tectonics has arisen that is not that much different from the old one, but it has the advantage of its starting impetus and of its intrinsic vigor. In the kind of loaded mobilism that I have presented, the new tectonics is increased by all the true power of the old one and can face, with all forces combined, the challenge of the known facts, while simultaneously opening, in the unknown of yesterday, extensive and fruitful alleys.

But mobilism, if it is true, occupies in the domain of dimensions, of forces, and of realizations a completely different place from the one it holds in the abstract domain. It actually controls the energetic distributions on the upper scale and on all scales.

Almost nothing is known about the forces responsible for continental drift, but one should admit, in addition to a passive transportation of the sial by the currents of the sima, movements proper to the former with respect to the latter. I have said how it is possible to visualize, at least qualitatively, the powerful control of basement folding and of the great traction effects. Almost everything else depends on the control of basement folding. The geosynclinal chains maintain their deformation by a frontal drawing of the energy of the continental block to which they are attached, at least in the lower part. To this is added a basal drawing from the energy conveyed by the sima. The geosynclinal chains, leaning more and more against the major part of the continental block, restitute to it, in the upper parts, at the margin of the jaws, a small fraction of the energy that animates them, and this fraction is consumed in basement folds of moderate tonnage, when it does not simply exhaust itself in brittle behaviors. The weak drawing of energy, (327) necessary for the generation of cover folds, is accomplished either directly from the energy of basement folding or from the energy transmitted horizontally, across sections of variable width of the cover itself, by the geosynclinal chains.

Thus, the energy is distributed, on the upper scales, in massive flows that divide themselves, on the lower scales, in streams and in increasingly smaller rivulets. Naturally, these visual metaphors do not clarify a physical theory of the distribution of energy. But correct visions can, if necessary, do without theoretical ballast; whereas the theories cannot succeed without a correct initial vision.

I shall not discuss the distributions of energy related to infratectonic objects.

Obviously, if the energy of drift is the main question, nobody will think of disputing the part played by numerous other sources of which I shall not make an inventory. Gravity, under a thousand direct or indirect forms, plays a fundamental role in these deformable media, which are incapable of supporting their own weight as soon as the column of matter reaches a certain thickness. It is superfluous to cite all the deformations attributed to isostasy. The lateral push derived from the filling, from the accumulation of the deposits in a furrow that subsides, can play a certain role in cover foldings of a certain thickness, as it does in incipient and not too broad geosynclines. The changes of volume due to recrystallization, and those generated by the evolution of deep-seated magmas, cannot be correctly estimated at present, but it would be dangerous to underestimate them. If the theory of contraction appears irreconcilable with mobilism, this is not the case for contraction itself, if it occurs. But, as has been pointed out, nobody knows at present whether the Earth is heating or cooling.

I shall never sufficiently stress what geology owes to the fruitful concept of *filling,* the apex of the thinking of Elie de Beaumont, which includes, clearly expressed or in a strongly implied form, most of the ideas with which tectonics has lived for a long time and with which it will always live, as long as the use of the concept is precisely regulated: the idea of framed folding, the idea of geosyncline, the idea of double chain and of double overturning, the ideas of unilateral overturning, of true foredeep, and of foreland. Those who have fought this concept owe the most to it. Is it not by eliminating from the concept of filling the lately accepted implication of rigid geometrics or of somewhat hard ossification that all its living substance has been unraveled? By means of such a game, fixist tectonics has gained a new half century of greatness and its richest blooming. But it seems today as if it tried to escape, without succeeding, from the difficulties it has itself raised. If the mobilist approach had helped only to open new horizons and to generate new checkings, it would have provided (328) a beautiful career for itself. At least it does answer one of the requirements of our time. Unburdened by several accessories that I did not mention and that few geologists would be willing to accept, the mobilistic approach adequately relates all the facts previously known with those I have discovered. I am not so naive as to say that it is the only theory that can achieve this relation: but I intend until now to maintain the superiority of concrete tectonics and to preserve the nuance that distinguishes it from theories of all kind.

With respect to my concept of framed folding on all scales, it seems to me that it reaches, as nearly as it is possible today, the perception of the total deforming movement I was discussing at the beginning. Changed into a notion and unfolded into ideas, this concept reaches very far into the background that is shared by the most diverse theories; it appears particularly capable of coordinating their compatible elements. Furthermore, this concept, because it is not in its essence a theory or an extract of theories and because it goes beyond the presently known forms of mobilism and fixism, is not necessarily related to any of them. This is a point that should be stressed. This concept is independent of theories with which it may temporarily display some relationships. It could not accept all of them or a given one entirely. Such an extreme laxity would be but weakness of judgment. But, it is capable of considering theories as working hypotheses only, of balancing one against the other, of bringing them together in useful contact, and, in this manner, of tempering their excesses, of rejecting their dead portions, and of associating their durable elements. By virtue of its great flexibility and the almost unlimited richness of its imagery, this concept has a chance to last if in its subsequent developments its aspect of live and flashing thought is preserved, and also if it is protected from any systematic induration.

A concept that leads to the predominance, on all scales, of the notion of framed folding, with all the mechanical procedures it implies, and that depends, on the planetary scale, on the same fundamental principle common to classical tectonics and to mobilism, such a concept may still appear schematical. But reattributing to this concept all that can be seen of the ceaseless deformations that have animated the terrestrial globe, it is difficult to believe that this concept may be so exceedingly pure. By the very condensed form under which it is capable of accounting, in its light contexture, for a whole world of ordered movements, it is clear and practical: this is the least I wanted to be acknowledged in it.

Such are, as I sense them, some of the possible arrangements of ideas, and such are the visions that are offered. In my presentation, I have attempted to equilibrate one with the other in order to avoid, as would be proper to a working attitude, the extremes of a naive dogmatism and of an impotent skepticism, while keeping in the deepest part of my mind a certain detachment as to the fate of theories or, better yet, as to their perishable element. (329)

The universe flows, carrying with it milky ways and worlds, Gondwanas and Eurasias, inconsistent visions and clumsy systems. But the good conceptual models, these *serena templa* of intelligence on which several masters have worked, never disappear entirely. They are the great legacy of the past. They linger under more and more harmonious forms and actually never cease to grow. They bring solace by the great art that is inseparable from them. Their permanence relies on the immortal poetry of truth, of the truth that is given to us in minute amounts, foretelling an order whose majesty dominates time.[37]

Epilogue: Asia

What else should I say? We have questioned all of Asia, and she has responded rather generously; she has informed us of other lands, and there are few she does not help us to understand better. We have reached in the end the Japanese islands, which are nobly curved and as if bent over the secret of the waters. Let us rest in these well-built lands where each morning the rising sun begins to light up Eurasia. The Fuji at dawn announces the glory of the day to come. From the depths of the blue immensity, waves rise, break, and thunder: they tell of the beautiful fugacity of appearances, of the measured equilibrium of things. Under our feet, less agile waves crowd themselves in the black depths. Far away, behind us, as far as the heart of the continent, other and still other waves, exhausted by time, congealed in the splendid torpor of the old chains, are reanimated through the immense efforts of the heavy basement waves. This is how in the course of time wavering veils concealed the old heart of the world. The waves pass and as in the old dreams of Asia they all together tell of the evanescence of the universe. How many times did the sun shine, how many times did the wind howl over the desolate tundras, over the bleak immensity of the Siberian taigas, over the brown deserts where the Earth's salt shines, over the high peaks capped with silver, over the shivering jungles, over the undulating forests of the tropics! Day after day, through infinite time, the scenery has changed in imperceptible features. Let us smile at the illusion of eternity that appears in these things, and while so many temporary aspects fade away, let us listen to the ancient hymn, the spectacular song of the seas, that has saluted so many chains rising to the light.

Illustrations

Figures 1 to 4. **Virgations***

Figure 1. **Simple virgation of the first type.**
Figure 2. **Double virgation of the first type.**
Figure 3. **Simple virgation of the second type.**
Figure 4. **Double virgation of the second type.**

The horizontal projection of the flow lines is shown by dashed lines. The use of such a graphic representation for these flow lines, which obviously should be represented by continuous lines, has been dictated by the necessity of not overcrowding the sketches. The arrows related to these lines indicate the direction of flow.

The density of the pattern of the flow lines in the central segment of the virgations of the first type emphasizes the fact that most of the flow and most of the energy of deformation are concentrated in that segment. The gradual decrease of these factors toward the wing or wings is expressed by the divergence of the flow lines and by the lesser density of the pattern.

In Figures 2 and 4, the central transverse alignment of the virgation, or the line of the easiest and most rapid flow, is shown by the dash-dot line.

The segments of obstacle whose margin determines the virgations of the second type are shown by oblique hachures (Figures 3 and 4). The direction of the stresses that stretch, longitudinally, the wings of these virgations is given as solid lines to which are associated arrows indicating the direction of this flow subordinated to the major behavior.

* See pages 37 to 43.

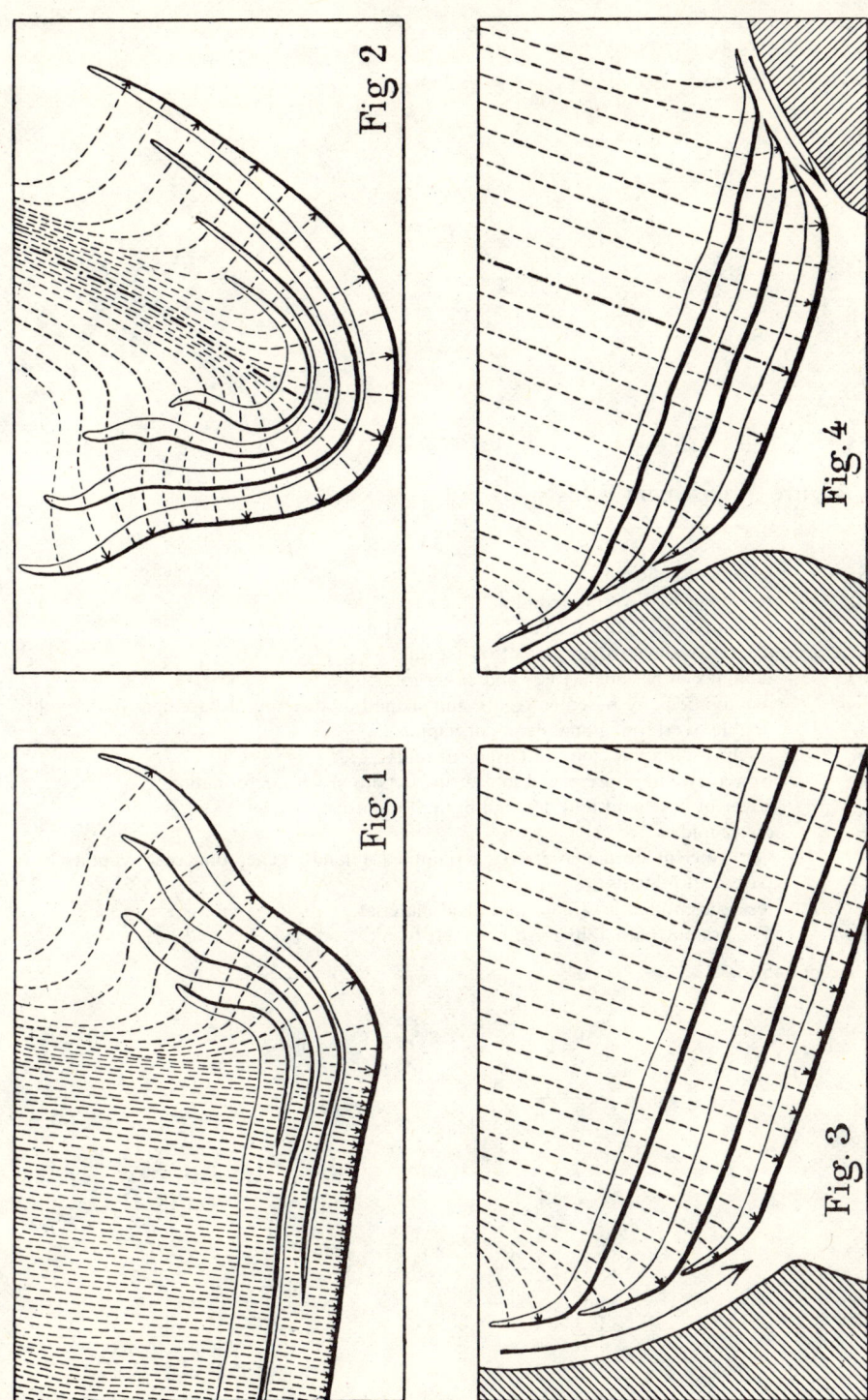

Figure 5. **Basement Folds***

A:	basement anticline.
B:	basement fold with longitudinal fractures.
C, D, E:	gradation of basement folds into clean-cut nappes.
At C:	frontal swelling (b) and latent thrust (c).
FF:	complex of basement folds and cover folds.
GG:	reactivations by basement folds; fan-shaped association of basement folds with double overturning and clean-cut nappes.
a:	fractures and voussoirs of basement folds.
b':	frontal swellings preserved in a thrust condition with deformation.
d:	cover of basement fold dragged in the thrust nappes.
e:	cover folds.
f:	very ancient dead material (reactivated foreland), generating reactivations by basement folds in *g*.
g:	less ancient and less indurated dead material.
Note:	The granitic batholiths have been left blank.

* See pp. 46 to 53.

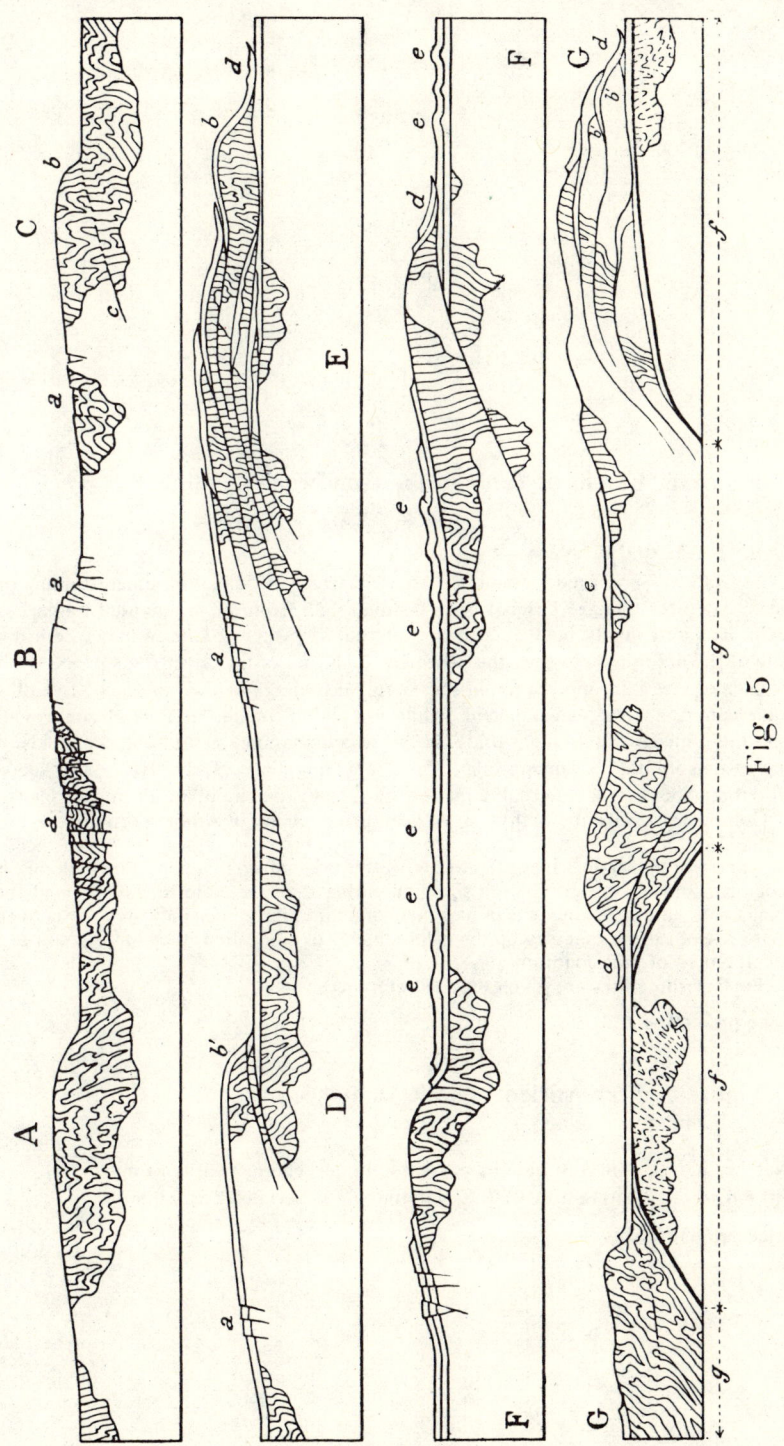

Fig. 5

Figure 6. **Basement Folds of the Gondwana Continent**†

Scale 1 : 120,000,000.

Legend. 1: Sima predominant. 2: Areas in which anticlinal basement folding predominates. The new tonnage has not been distinguished from the reactivated tonnage. On this scale, it is completely negligible in the internal virgation of Gondwana, where it consists only of the deformed cover of the basement folds and of the cover folds properly speaking. It reaches a certain importance in the marginal foldings of the continent, from the South American Andes to New Zealand, while remaining, in this peripheral zone much lower than the tonnage of basement folds. 3: Axial culminations of the basement folds. 4: Axial depressions of the basement folds. 5: Connecting lines. I, II, III: First, second, and third branches of the internal virgation of Gondwana. a, b, c: Promontories displayed by Gondwana toward the Tethys. a: African promontory, b: Arabian promontory, c: Indian promontory.

Although the figure, in order to show the original connections and linkings between basement folds, corresponds to a situation earlier than the major disjunctions of the Gondwana continent, the main areas of basement anticlines have been indicated, by convention, in their present extent, that is, with the enlargements or elongations they have acquired after the development of that condition.

Present shorelines are given only as references.

† See pp. 149 to 157.

Figure 7. **Very Ancient Asiatic Drifts**‡

Sketches a, b, c: Three situations presented in descending order of time.
Legend. 1: Siberian massif. 2: Sinian massif. 3: Serindian massif.

‡ See pp. 83 to 85.

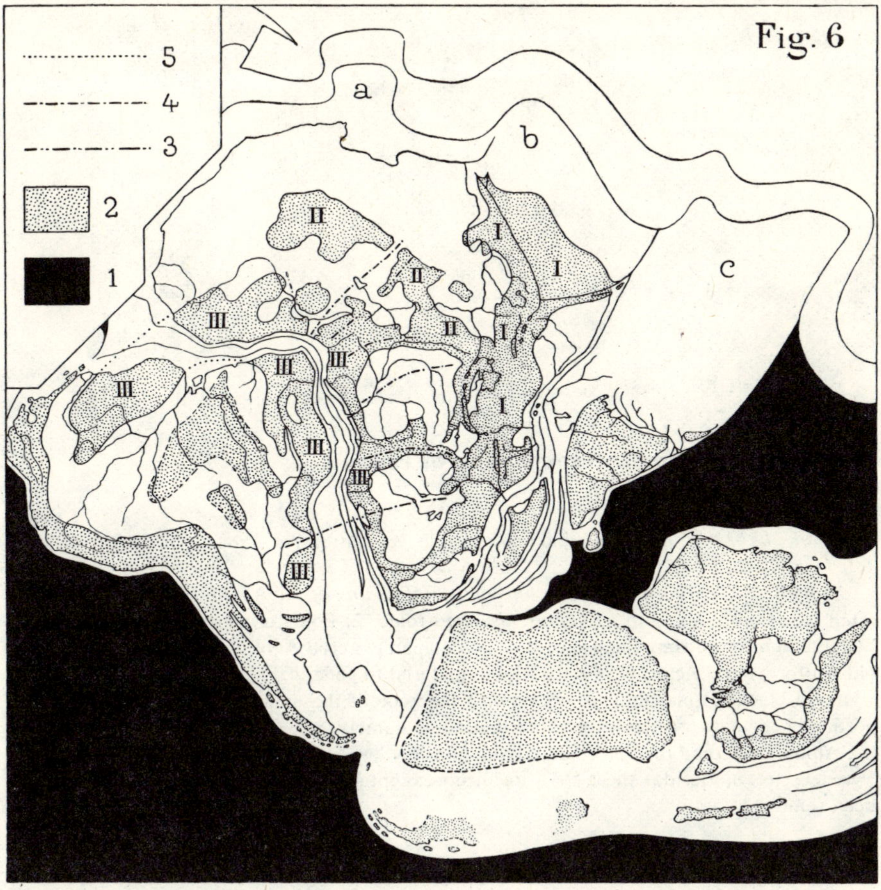

Illustrations 171

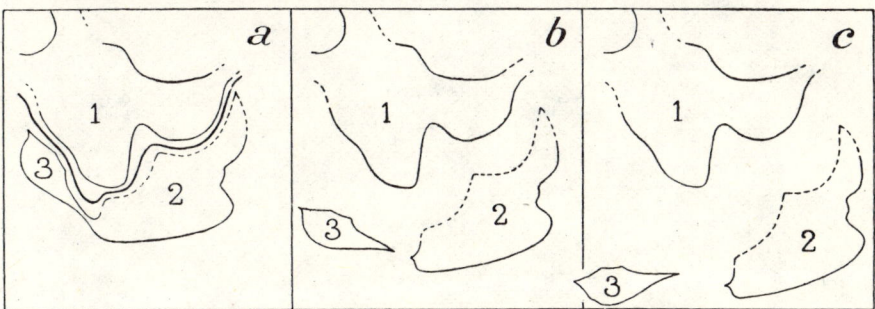

Fig.7

Figure 8. **Schematic tectonic Map of Eurasia**

Scale of the main map. 1:60,000,000. *Scale of the insert showing Southeast Asia.* *1:120,000,000.*

Legend. 1: *Geosynclinal chains and marginal chains of the Alpine cycle,* with incorporated basement folds. 2 to 6. *Alpine basement folds.* 2: Folds of Hercynian material; 3: of material of inferred Hercynian age; 4: of Caledonian material; 5: of Precambrian material and old platforms in general; 6: of ante-Alpine material in general.

Solid black (1) indicates, in addition to the zone of the main new chains, a very small portion of the new tonnage external to that zone, namely, a few important cover foldings of the Alpine cycle and the deformed covers of a few large Alpine basement folds (Caucasus, Pyrenees, etc.). Besides these very localized exceptions, the covers are assumed to have been removed.

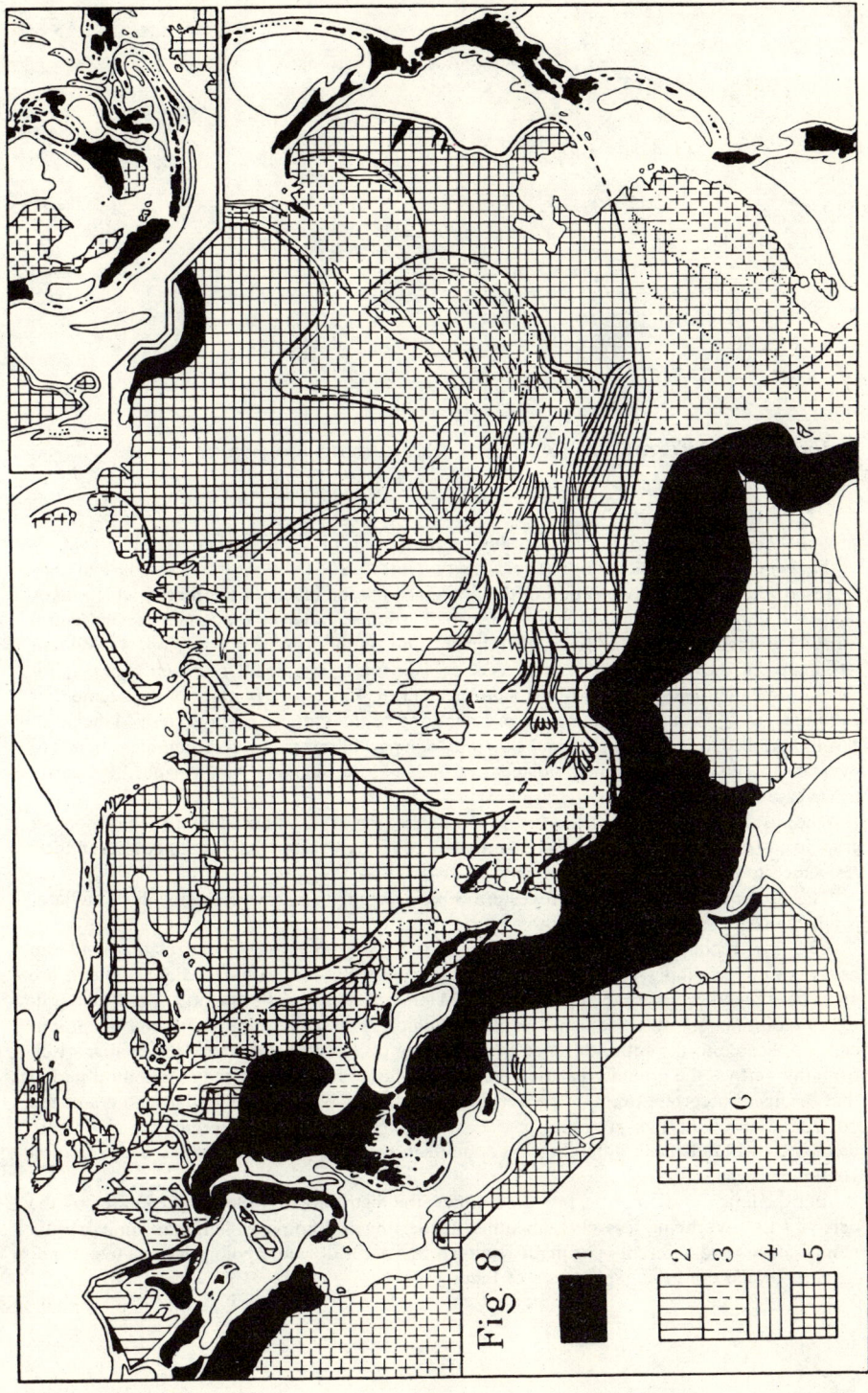

Figure 9. Deformation Regime of Asia during the Alpine Cycle

Scale of the main map and of the two inserts pertaining to northern Asia. 1:45,000,000.
 Legend. a, b, c: Connections between the two inserts and the main map. The dashed lines represent the horizontal projection of the flow lines for the levels of the flux in which takes place the command of the visible foldings. d: Particular transverse alignments. 3: Central transverse alignment of the Turanian segment and of the Turanian virgation. 4: Western terminator of the segment of central Asia. 5: Transverse alignment common to the Indo-Siberian segment and the Indo-Mongolian segment. 6: Eastern terminator of the segment of central Asia. 7: Transverse alignment of Lena-loop of the Patoms-northwestern Manchuria. The flux of the Greater Khingan, in the vicinity of a part of this transverse alignment, tends to override the flux of the segment adjacent to the east, the segment Aldan-Amur. 8: Central transverse alignment of the latter segment.
 The distribution of flux inside the most indurated massifs is not shown. For reasons of graphic convenience, these massifs remain blank, but it is obvious that the superiority of their resistance to plastic deformation is very relative.
 The center of the depression of western Siberia, from which divergent flow lines radiate, is left blank for other graphic reasons.
 The flow regime being not independent of time, a graphic expression of this kind can apply, strictly speaking, to a single instant of time only. But for old Asia, whose elements are all welded together before the opening of the Alpine cycle and subsequently operate according to an intimately synergic behavior, the graphic expression can be considered as practically applicable to the entire duration of the Alpine cycle. With respect to the chains arisen from the Tethys, the graphic representation of the flow lines is practically valid for a period that begins at a certain point in the Tertiary paroxysms, when the essential features of the confrontation of the two jaws and of the arrangement of the new intermediate chains are developed without much difference in comparison with the present situation. This duration continues today.
 For the long duration of the precursor times, the distribution of the flux, regulated by the behavior of jaws being less close together at the times of compression, was similar in its main features and different in its main details; it was naturally undergoing changes over time.
 See, in addition, the explanation of Figure 12.

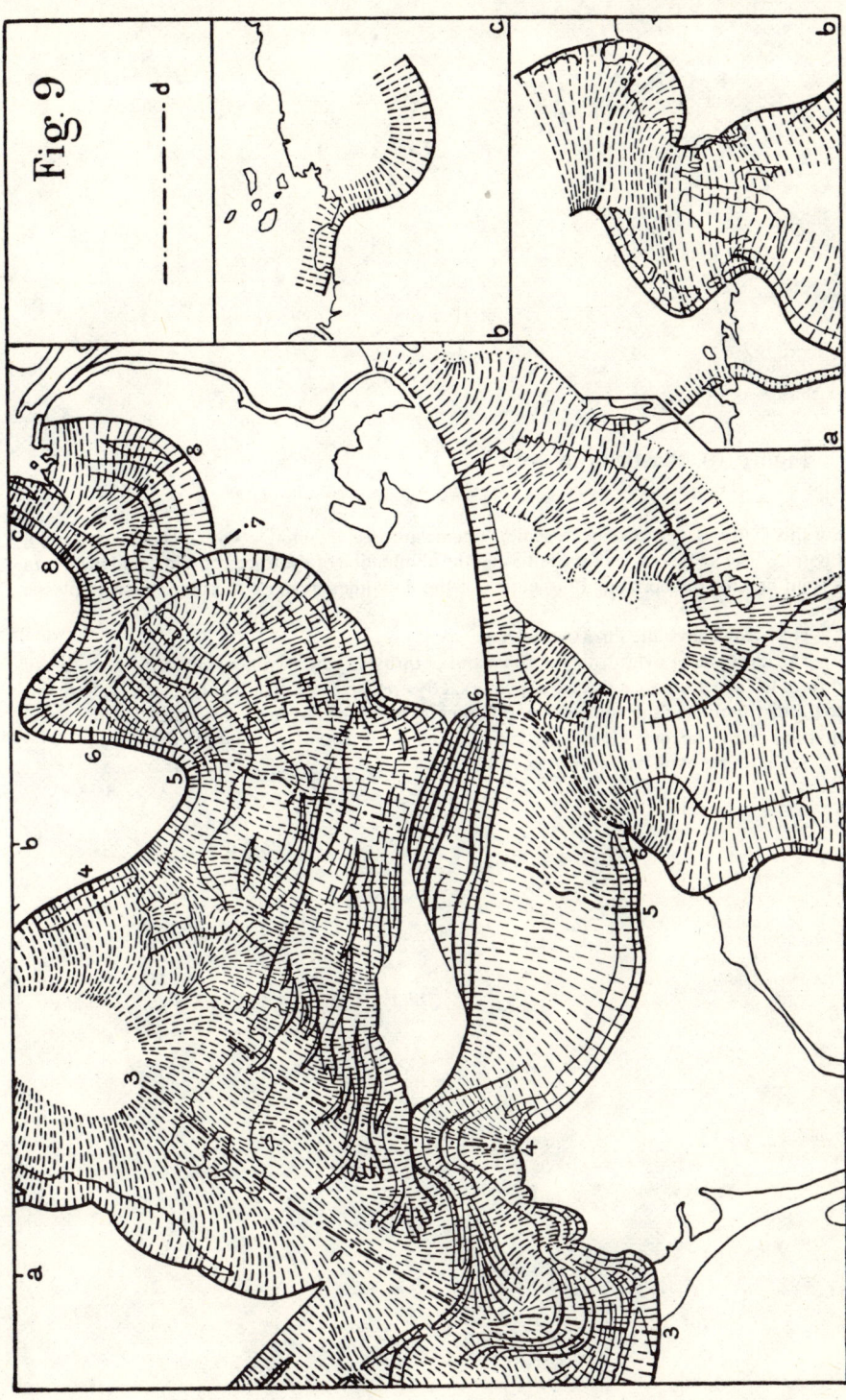

Figure 10. **Plasticity of Asia**

For this figure, the arrangement, scale, nomenclature, and comments are identical to those of Figure 9. The suppression of the lettering, the elimination of the traces of folds, and the use of a solid line for the horizontal projection of the flow lines increase the impression of plastic flow.

The arrows give the direction of flow.

The direction of overturning of folds and of thrusts, generally identical to the direction of flow of the general flux, may also appear exactly opposed to it in particular cases that are due to the distribution of the velocities at different levels of the flux and, consequently, to the fact that the matter becomes internally complicated by apparently retrograde folds.

See also the explanation of Figure 12.

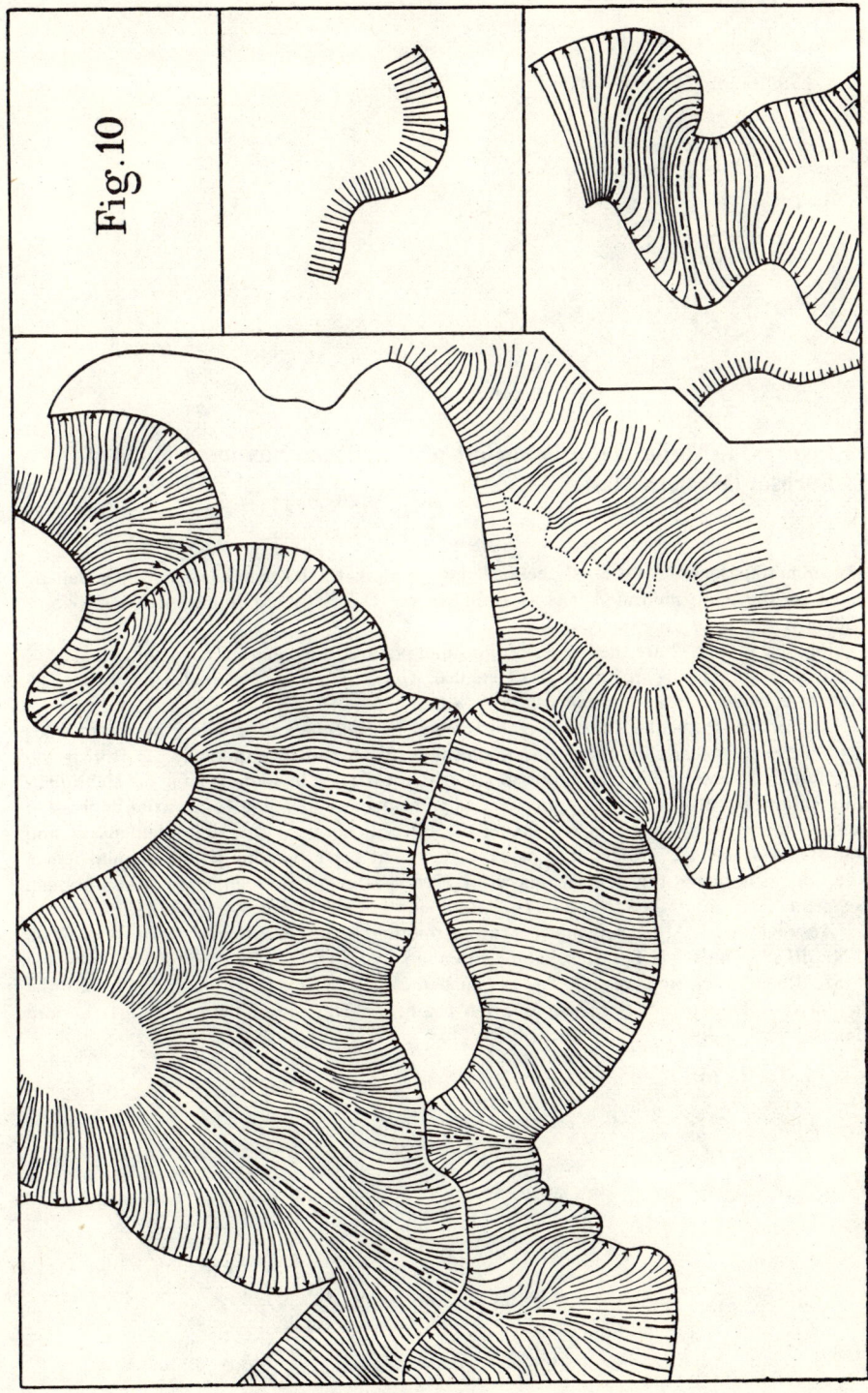

Fig. 10

Figure 11. **Major Axial Behaviors of the Alpine Basement Folds of Eurasia (Western Half)***

The numerous longitudinal cross-sections show the great axial behaviors and their dependence on the more indurated massifs facing them. The latter are represented by oblique hachures.

Figures 11 and 12 are the halves of the same picture; they share a connecting band.

Approximate horizontal scale. 1:30,000,000. *Approximate vertical scale.* 1:750,000. The curvature of the Earth is not taken into account.

Legend. a: Particular transverse alignments. b: Designation of a few basement folds whose direction leads to their being intersected other than longitudinally. 1: British and Lusitanian transverse alignment corresponding to the axial culmination facing the Hebridian-Laurentian massif. 2: Parisian transverse alignment corresponding to an axial depression and to the easiest flow direction facing the interval between the Hebridian massif and the Baltic massif. 3: Armenian transverse alignment corresponding to the culmination of the Caucasus facing the Arabian massif. 4: Central transverse alignment of the Turanian segment.

Abbreviations. Ard.: Ardennes. Arm.: Armorica. Cant.: Cantabria. Corn.: Cornwall. FN.: Black Forest. H.L.: Hebridian-Laurentian masif. MC: Massif Central of France. MSR: Rhenish schistose massif. P.f. arc anat. mér.: Basement folds incorporated into the southern Anatolian arc. Sc.: Scandinavian chain. T.: Thuringian Forest. Tb.: Teutoburg Forest. V.: Vosges.

* See pp. 90 to 101.

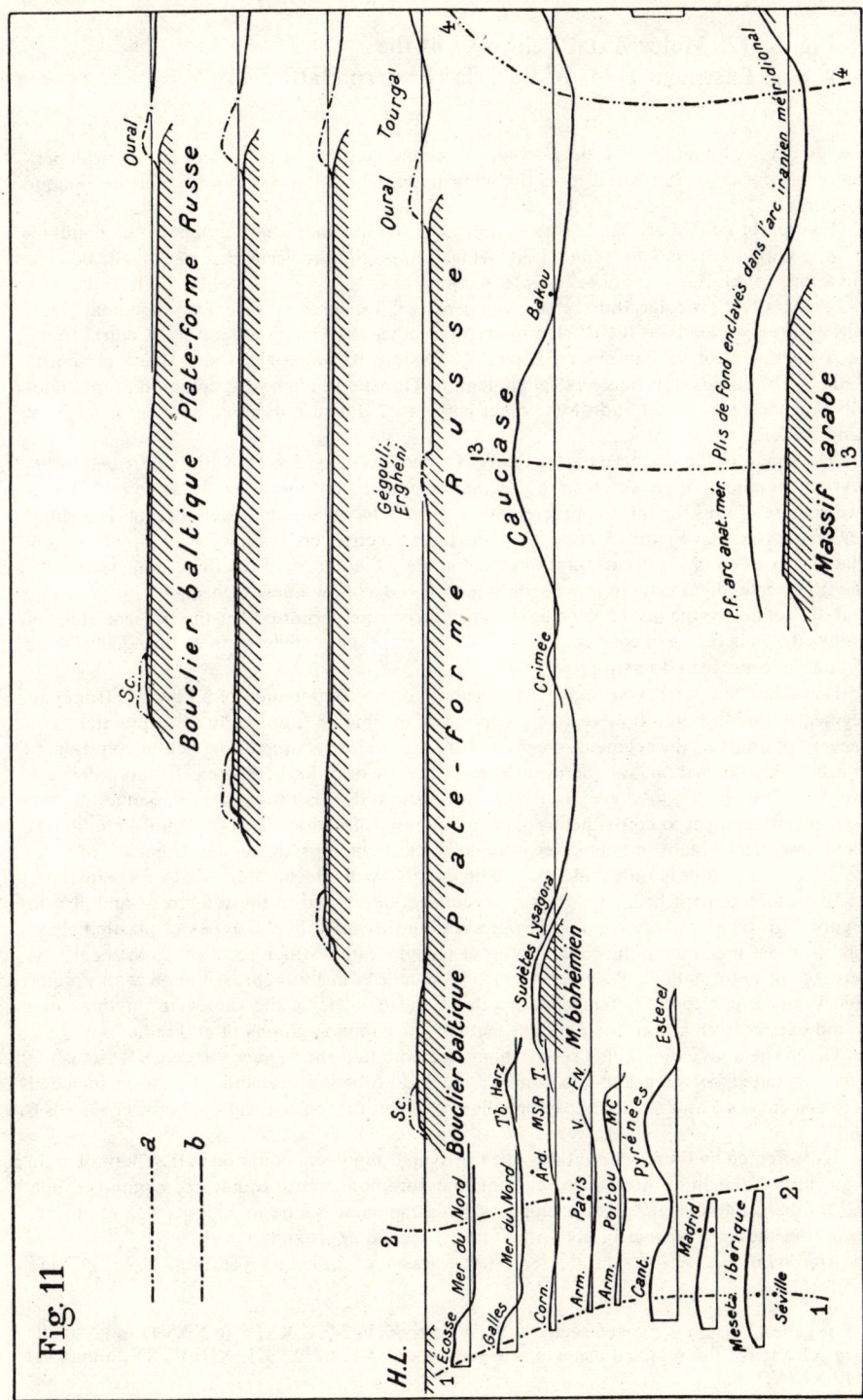

Fig. 11

Figure 12. **Major Axial Behaviors of the Alpine Basement Folds of Eurasia (Eastern Half)***

The numerous longitudinal cross-sections show the great axial behaviors and their dependence on the more indurated massifs facing them. The latter are represented by oblique hachures.

Figures 11 and 12 are the halves of the same picture; they share a connecting band.

Approximate horizontal scale. 1:30,000,000. *Approximate vertical scale.* 1:750,000. The curvature of the Earth is not taken into account.

Legend. a: Particular transverse alignments. b: Designation of a few basement folds whose direction leads to their being intersected other than longitudinally. 4: Central transverse alignment of the Turanian segment. 5: Western terminator of the segment of central Asia. 5a: Irkutsk-Hami transverse alignment. 6: Transverse alignment common to the Indo-Siberian segment and the Indo-Mongolian segment. 7: Eastern terminator of the segment of central Asia.

The shape of the northern parts of curves 5 and 7 is in agreement with the content of the text and similarly originates from a *practical* simplification conceived so as to include between these curves the most characteristic vertical behaviors (axial behaviors) of the central region of Asia. But Figures 9 and 10, although planimetric, originate from a graphic concept that allows us to take into account more completely the effects of the three dimensions. All the transverse alignments in it are strictly conceived as flow lines. Consequently, Figures 9 and 10 define by means of flow lines, as precisely as permitted by the present state of knowledge, all the segments of flux and all the transverse alignments of flux conceivable within the represented territory.

It is clear that, strictly speaking, there should be no continuation of a flow line from one segment to another in a flow regime independent of time or supposedly so in practice. The concept of limits of the segment of central Asia is therefore a simplifying fiction convenient, in a first approximation, for the discussion of the major axial behaviors. In more delicate problems, in which a greater approximation is required, it is necessary to assume that the domain with respect to flow remains open in these two directions. Figures 9 and 10 fulfill this requirement and many others: they represent the behaviors of all the segments of flux, including their major features and their main details, with a higher degree of approximation.

In a more general fashion, the graphic concept expressed in these figures—and also in Figure 20B—is extremely capable of representing the most diverse types of plastic behaviors. Without mentioning the capability of visual suggestion which reaches, so to speak, the very act of deformation, the graphic concept succeeds in this representation with an efficiency and a precision by far superior—the initial data being the same—to anything one would expect from a text, longitudinal sections, or common graphs of all kinds.

Given the above-mentioned reservations, we note that the segment of central Asia is the space included between curves 5 and 7; the Indo-Siberian segment, the space included between curves 5 and 6; the Indo-Mongolian segment, the space included between curves 6 and 7.

Not affected by these reservations, the transverse alignment common to the Indo-Siberian segment and the Indo-Mongolian segment, the Turanian central transverse alignment, and the southern parts of the two terminators have the same shape in Figures 12, 10, and 9; therefore, they are correct in both a first and a second approximation.

Abbreviations. A: Assam. K: Kurla. L: Basin of Lukchun (Turfan), P.R.: Russian platform.

---

* For general discussions, see sections VI, VIII to XIV, XXI, XXIV to XXVI, and XXVIII to XXXI; for the regional aspects, see sections V, VII to IX, XI, XIII to XX, and XXII to XXIV.

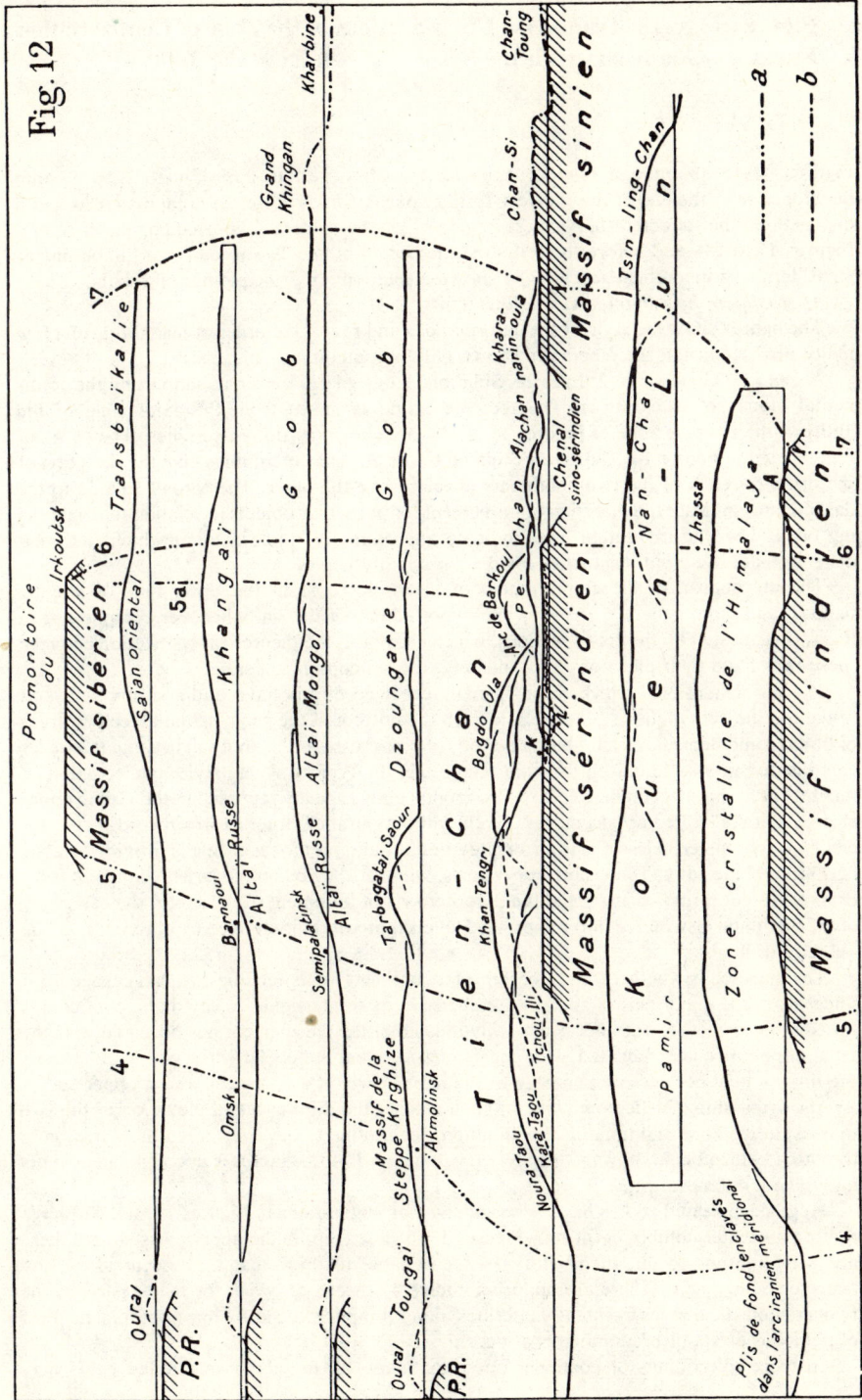

Fig. 12

Figures 13 to 18. Transverse Cross-Sections of the Zone of Confrontation Eurasia—Gondwana with the System Arisen from the Tethys

Figures 13 to 18 build a homogeneous series. They pertain approximately to a state developed near the end of the greatest Tertiary paroxysms, before the great distensions and disjunctions that affected the western segments of the system—the Mediterranean frame, Figures 15 to 18—and before the drifts over lenses. It is needless to add that the outline at great depths is theoretical and aims at showing solutions of principle not of details.

Approximate horizontal scale. 1:10,000,000.

The figures show the style of the deformations and reveal the order of magnitude of a few major displacements; the others are shown only in a qualitative manner.

Legend. 1: Gondwana. 2: Eurasia. Solid black designates the sima, supporting the continental blocks of sial (blank). The tectonic products arisen from the axial zone of the Tethys—the Pennine zone of Figures 15 to 18—are shown by dotted surfaces. They are not represented in Figure 14. The arrows related to the dashed-dotted lines give the direction of the drifts for each of the two continents in relation to the other. The arrows related to the dashed lines indicate the direction of movement for particular objects in relation to neighboring ones. The vertical dotted lines indicate approximately the places at which the great distensions or the great disjunctions will subsequently begin.

The mechanism of the emplacement of the basic rocks at the lower face of moving nappes, and more generally between the major eddies of the sial whenever complicated in itself, is suggested by the cross-sections, which show also the theoretical connection between these rocks and their place of origin, namely, the subcontinental sima.

The old distensions, which through many renewed efforts have outlined, created, and reworked the Tethys, have generated through the bottom of the geosyncline extensive areas of thinned and depressed sial. During periods of compression, this thin sial became internally complicated in a style that in its main features is the typical plastic style, namely, that of recumbent folds. A recumbent fold is a horizontal eddy, usually recorded in the visible upper part by parallel rolled-up structures. Even without parallel structures to record it, the recumbent fold nevertheless exists as an eddy in the sial, and the recording is provided by the periphery of the eddy. The visible upper parts, whether they belong to large recumbent folds or to clean-cut thrusts that extend the deformations of basement folds, reveal the capability of the flux to be distributed into superposed levels that are characterized by differences in the velocity of flow.

The question is whether the major part of a continent, in its total height, can engage itself underneath the major part of another continent, in its total height. To say that a continent is thrusted over another does not necessarily mean that the engagement has occurred with the great importance just indicated. This engagement, no matter how large the visible or admissible thrusts may be, may pertain only to the upper parts of a continent whose upper eddies overflow the antagonistic continent. In such a case, the differences of elevation of the two masses, in the zone and time of confrontation, seem often to have regulated the direction of thrusting, which occurs toward the less elevated jaw. But this factor is not alone and events may take a different turn.

Huge deep-seated eddies have been indicated as suggestions in Figures 13 and 14. I have drafted a smaller number of them in Figures 15 to 18, except in the upper parts, where there are good reasons to put them. But, it is reasonable to think that they occur along any transverse alignment. There are opposing eddies by means of which the major parts of the two antagonistic masses eventually interlock along complicated suture lines: this plastic form of behavior exists or predominates at depth.

In these overridings of continent upon continent, all the above-mentioned cases may

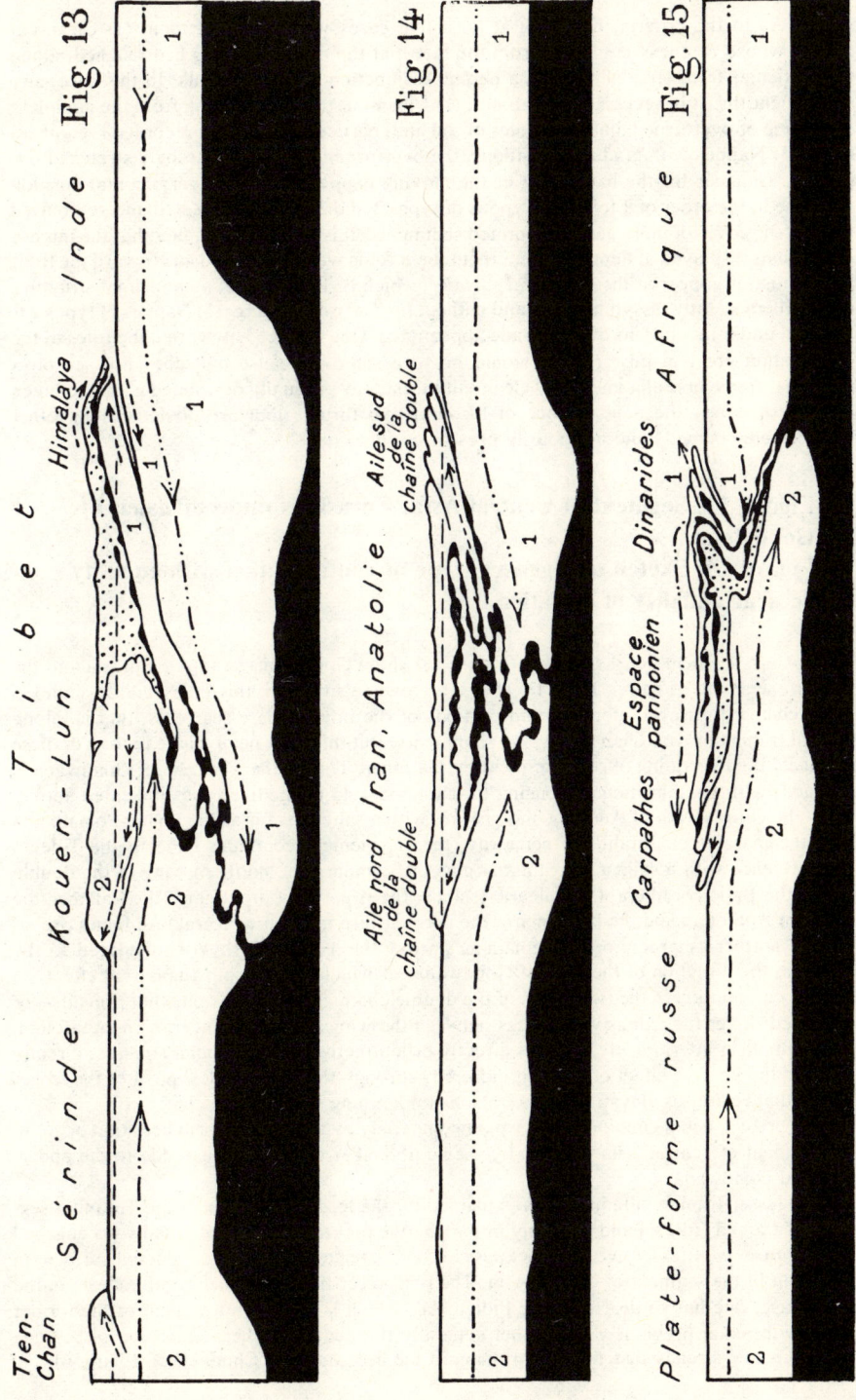

occur. As to the criteria, there are at least five cases worthy of attention. (1) Clean-cut thrusts whose neatness originates from the fact that the old distensions had reached, along the particular transverse alignment, a perfect disjunction of the two sials. In this case, any subsequent thrust is necessarily clean-cut. (2) Clean-cut thrusts resulting from the complete stretching of overturned limbs of eddies of sial and, particularly, of well-recorded recumbent folds. (3) Nappes with gradual transitions, the overturned limb being strongly stretched but present. Chances for the basal sima of the moving nappes to reach a very frontal position decrease in the order of 1 to 3. (4) Nappes of types 2 and 3 but little disected and still buried under their cover of more gently deformed sediments. It is a well-known fact that the intense contortions displayed at depth by great recumbent folds weaken upward and toward the front and almost disappear at the *structural surface*, which is, in principle, an anticlinal structure of cordillera slightly dissymmetrical and without *the last overturning*. (5) Nappes of types 1 to 3 buried under involutions of retrograde appearance. One can see how varied the rules of the deformation are and how rash it would be to assume—because one does not see on a particular transverse alignment any clean-cut thrust, any gradually developed nappe, or even any overturning—the nonexistence of bicontinental thrusts that are obvious along other transverse alignments and necessarily present on that one.

Figure 13. Segment of Central Asia. Eurasia is underthrusted by Gondwana.

Figure 14. Sketch of a general type of confrontation with equality or near equality of elevation.

The eastern segments of the system (Figure 13) show Gondwana on top of Eurasia and the western segments (Figures 15 to 18) show the reverse situation; this intermediate situation has a chance to prevail along certain portions of the intermediate segments, that is, along several Iranian or Anatolian transverse alignments. But this does not exclude for any of these segments the overriding of one continent by the other. The northern margin of Gondwana is engaged, with complications belonging to one or several of the five types indicated above, along the entire width of Anatolia and of Iran, with a salient as far as the Pontic domain and near the margins of Turan. Farther away, the engagement continues beneath the Tibetan intumescence with a salient near the margins of Serindia. The northern wing of the double chain (the Betic cordillera, the Balearic Islands, the Alps, the Carpathians, the Balkans, the northern Anatolian and the Iranian arc, the northeastern margins of Karakorum, and probably the northern margins of the remaining part of the Tibetan Tethys) is displaced to the north, in the direction of the drift of Gondwana, of which it is the most apparent effect.

The culmination of the two wings of the double chain and the lesser one that prevails—or prevailed—over the extensive surfaces between the wings (the Tibetan, Iranian, Anatolian, and Pannonian interiors, etc.) results directly or indirectly from tangential stresses; directly through the vertical effect of folding; indirectly through the Archimedes' push of the lenses of sial that restitutes elevation to the regions undergoing erosion.

The great Tibetan culmination is explained not only by the strong resistances that occur in the segment of central Asia but also by the additional Archimedes' push due to the underthrusted Gondwana.

The lesser Iranian culmination is explained by the lesser resistance of the Turanian segment of the old Eurasia and probably in part by the fact that the sial of Gondwana engaged with Eurasia is of lesser average thickness and of lesser tonnage, given an equal surface, in Iran than in the segment of central Asia. The reason for this difference is quite clear: in the latter case, one has to deal with the Indian sial, which is strong and normal; in the former case, with a sial previously and rather generally thinned, a northern extension of the one located in the already distended depression of the Mozambique Channel (cf. Figure 6).

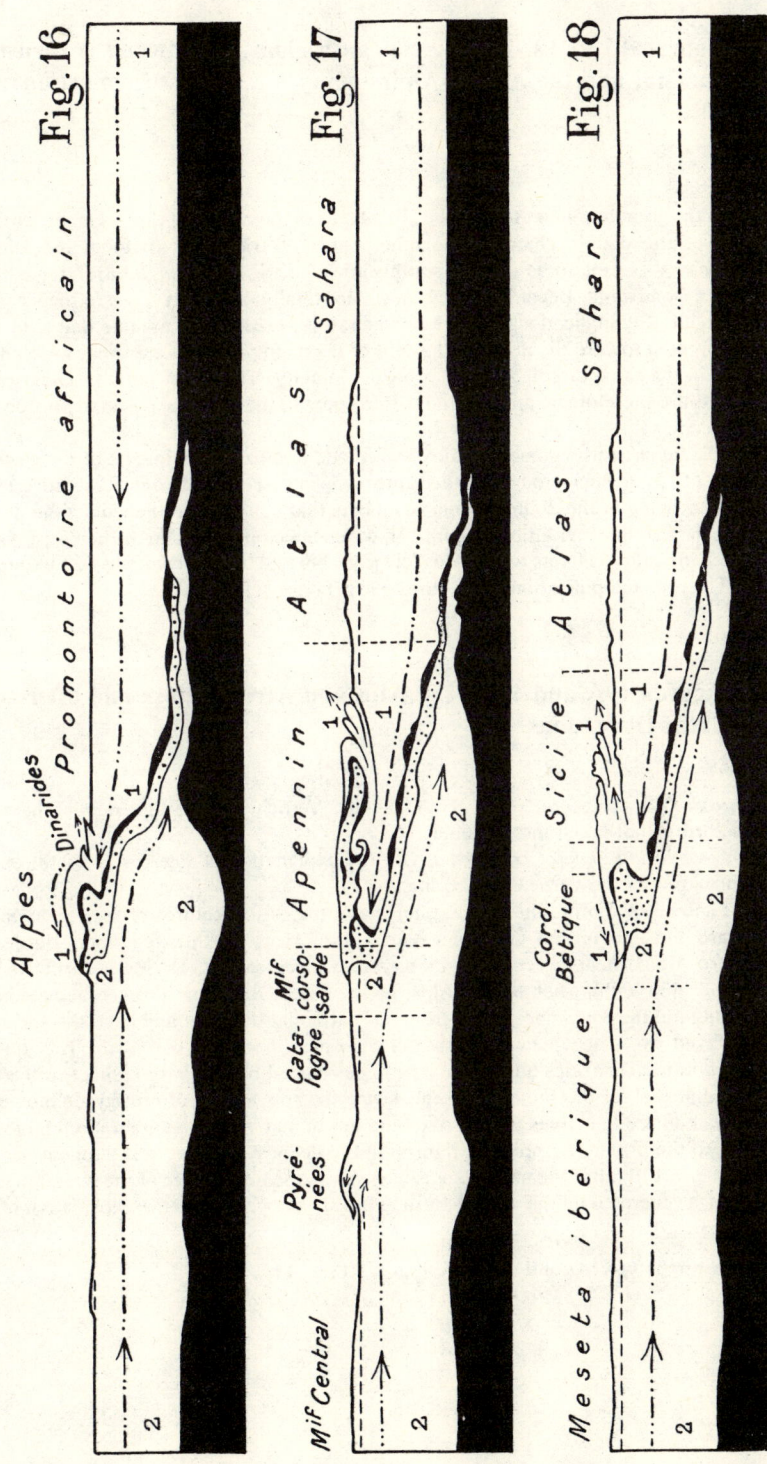

186   TECTONICS OF ASIA

Figures 15 to 18. **Mediterranean regions. Gondwana is thrusted upon Eurasia, and the Austro-Alpine nappes represent its most advanced salient.**

From the great basement fold of the Himalayas to the chains of the Atlas, the entire southern wing of the double chain—the southern arcs of Iran and Anatolia, the Dinarides, the Apennines—seems to be moving southward, which is true only with reference to the major part of Gondwana. Because these objects are totally or with respect to their essential parts the result of complications with Gondwana, it is necessary to assume that their movements are directed toward the north as the drift of the main part of Gondwana, which underthrusts them in the same direction but at a higher velocity. The upper parts of Gondwana in these objects are therefore lagging behind with respect to the main deep-seated portion of the same continent.

The underthrusts sometimes involve in addition to the sediments of the earlier northern slope of Gondwana—today molded into nappes or into folds over the entire length of the southern wing of the double chain—elements that belonged to the axial zone of the Tethys. Thus the Pennine involution (Figure 17), which envelops Africa in the northern Apennine and which in reality belongs to the Alps (cf. page 146 and the explanations of Figures 26 and 27).

The regional indications in Figures 13 to 18 are self-explanatory.

Figures 19A and 19B: **The Alps and Africa before and after the Great Distensions***

Approximate horizontal scale. 1:14,000,000. Vertical scale exaggerated. The curvature of the earth is not taken into account.

Legend and general comments. See the explanation of Figures 13 to 18, excluding the explanation of the vertical dotted lines.

Figure 19A (before the distensions) shows the major features of the condition developed toward the end of the Oligocene paroxysm. Africa is thrusted upon Europe, and the Austro-Alpine nappes represent its most advanced salient. During its drift Africa became complicated within, behind the Alps, giving rise to folds and thrusts facing south: the Dinarides and the Apennines result from this particular behavior and from this local tucking up displayed by Africa, which is essentially displaced toward the north. The upper parts of Africa in the Dinarides and in the Apennines lagged behind with respect to the main deep-seated portion of the same continent; hence the appearance of retrograde movements. The vertical dotted line gives the approximate position at which the great stretchings will begin, prior to the probable ripping of the sial in the deepest troughs of the Ionian Sea.

Figure 19B (after the main distensions) shows the importance of the stretchings and of the tearings generated by the drift of Europe toward the north (see the explanation of Figures 22 and 23).

* See pp. 140 to 147 and compare Figures 13 to 18.

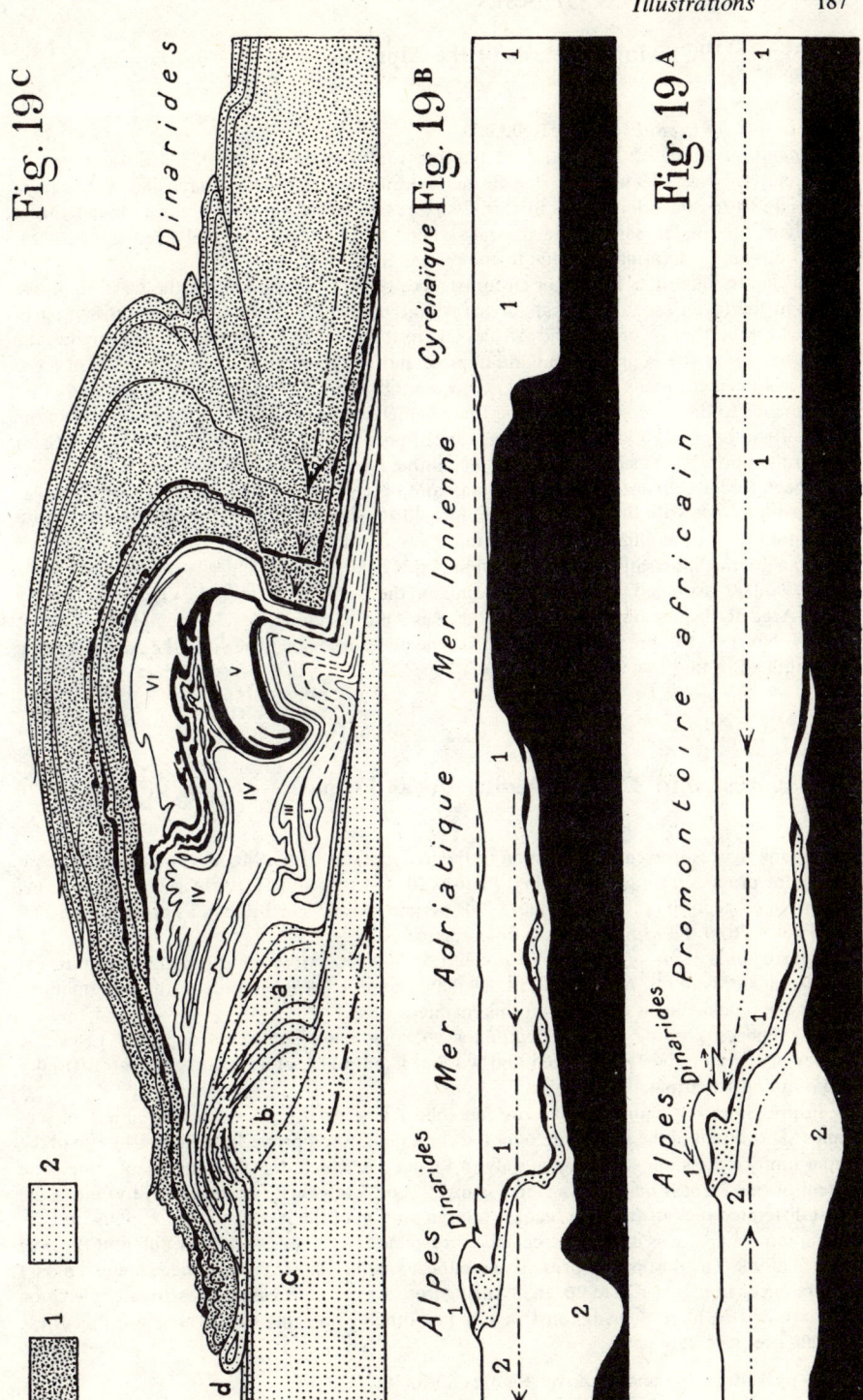

Figure 19C: **Major Details of the Alps**

Approximate horizontal scale. 1:1,000,000.

Legend. 1: Africa. 2: Eurasia. *a, b:* Basement folds of the margin of old Eurasia (on the first zone, see pages 93 and 94), due to the restituted energy, with decreasing effects from the inside to the outside (clean-cut thrusts expressed in the internal areas *a:* latent thrust underlying the major part of the frontal swelling *b*). *c:* More external deep-seated area, almost devoid of deformations due to the restituted energy.

This figure illustrates the major circumstances of the emplacement of the basic rocks (in black) at the lower face of several Austro-Alpine nappes, at the base of the entire Austro-Alpine system, that is, of thrusted Africa, and in the Mesozoic of the Piedmont furrow, the deepest basin of the Pennine zone and consequently of the Tethys. This last batch of basic rocks has been gradually rolled into the portion of the great Pennine recumbent folds that corresponds to the reversed limb of the Dent Blanche nappe (VI), the entire periphery of the Monte Rosa nappe (V), and the least advanced part of the normal limb of the Great Saint Bernard nappe (IV). Small intrusions of laccolithic type, fed from beneath by the great sills, have been forced upward into the anticlinal core of nappe VI. These processes of emplacement, which go on with the various vicissitudes during the Alpine precursor times, end before the last efforts of the Oligocene paroxysm.

For a graphic illustration of the evolution of this chain in thirteen cross-sections on a scale of 1:1,000,000, arranged in time succession from the Carboniferous to the Oligocene, see by Emile Argand the memoir entitled "Sur l'arc des Alpes Occidentales," *Eclogae Geol. Helv.*, Vol. 14, No. 1 (1916), pp. 145–191 (2 plates). The penetration of Africa into the Alpine system is mentioned in that memoir.

Figures 20 to 27. **The Mediterranean Frame***

These nine figures present a restitution of the major states of the Mediterranean frame from the Alpine precursor times until today. Figures 20, 21 to 25, and 27 form a series whose terms are arranged according to time. Figure 20B pertains to the same epoch as does Figure 20A and Figure 26 to the same epoch as does Figure 21.

Approximate scale of Figures 20 to 25. 1:55,000,000. *Approximate scale of Figure 26.* 1:18,000,000. *Scale of Figure 27.* 1:18,000,000. The order of magnitude of the deformations and of the displacements is given in this manner.

The figures provide with a degree of approximation sufficient for many tasks the framework, continuously deformed and without fixed point, of a redistributed stratigraphy and of a consistent paleogeography.

In principle, the figures show only the solid frame, that is, the essential object of tectonics. The extent of the seas that cover the continental shelf properly speaking—beyond the upper limit of the slopes—is shown only in Figure 27, which depicts the present state. The extent of the deepest marine areas in Figures 22 to 25 obviously corresponds to that of the most distended regions (in solid black). A large portion of the less distended regions (in gray on the same maps) was under the sea and corresponded to slopes with their different types of deformations. The portions of present shorelines, whether intact or deformed, are given only as references in Figures 20 to 26. In all the figures, the parts previously destroyed by erosion are assumed to have been reconstructed. For other comments concerning most of these figures, see page 196.

* See pp. 140 to 147 and compare Figures 15 to 19C.

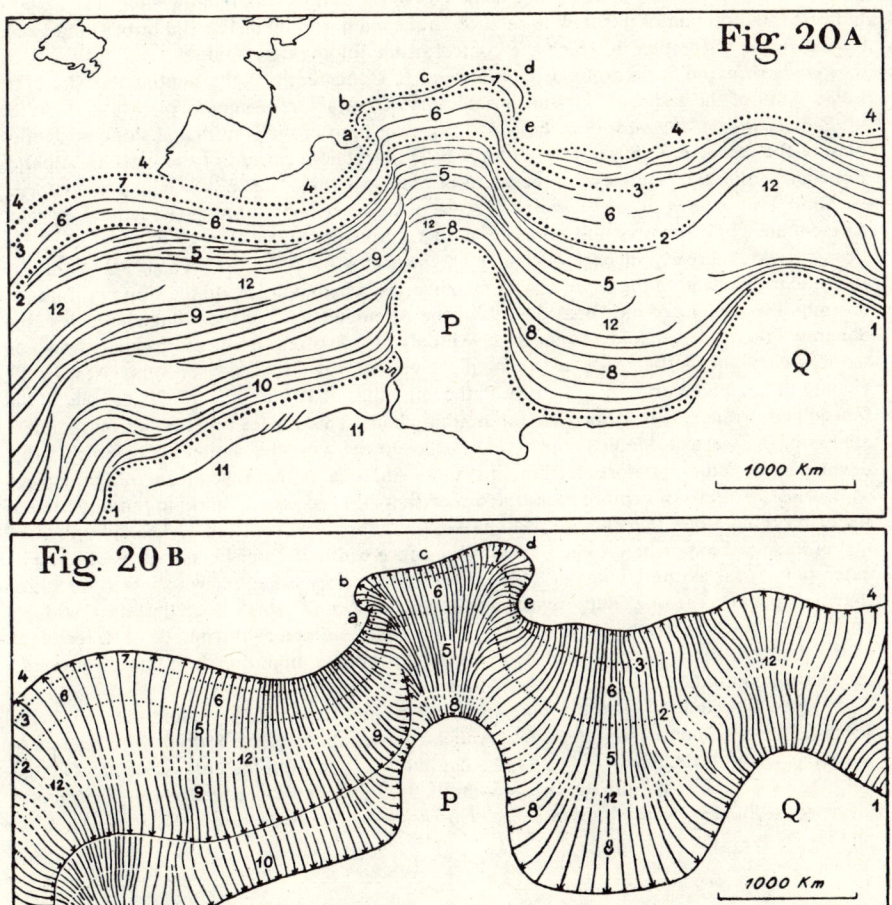

Figures 20A and 20B: General Features of the Mediterranean Frame during Alpine Precursor Times

Figures 20A and 20B show the major peculiarities of the old margins of both Gondwana and Eurasia. Figure 20A shows the general shape of the embryonic stages—furrows, cordilleras, or simple wrinkles, according to times and locations—of the chains of the Tethys. Figure 20B presents a corresponding state of the plastic flux of the Tethys with its flow lines. The arrows indicate the direction of this flow in relation to the main portion of Eurasia or of Gondwana, for the stages of the flux in which the control of the folding takes place.

Peculiarities of the margin of Gondwana. 1: Upper limit of the continental slope. 2: Lower limit of the same. P: African promontory. Q: Arabian promontory.

Peculiarities of the margin of Eurasia. 3: Lower limit of the continental slope. 4: Upper limit of the same. a: Ligurian promontory. b: Hemicyclic reentrant of the western Alps. c: Bohemian salient. d: Hemicyclic reentrant of the Carpathian region. e: Getic promontory.

Embryonic stages of new chains. 5: Austro-Alpine nappes forming the frontal margin, differentiated into furrows and cordilleras, of Gondwana. 6: Axial zone of the Tethys, or Pennine zone, already in part overriden in the south by 5. 7: Embryonic stages of the Helvetic nappes and of the lower Carpathian nappes, which complicate the slope of Eurasia. 8: Embryonic stages of the Dinarides; 9: of the Apennines; 10: of the Atlas chains less the Saharan Atlas, a cover folding that is shown only (11) to stress its future participation—on the left wing—in the behavior of the incipient virgations that already encompasses, in the vicinity of the main part of promontory P, the left wings of 10 and of 9. 12: Axial zone of the fan-shaped arrangement formed by the double chain. This object repeats—in an incipient stage—in the western Mediterranean and farther to the west the similar arrangements that develop in the future interiors of Tibet, of Iran, of Anatolia, of the Aegean, and of Pannonia.

During the times of compression, the direct flux—the flow to the north in the direction of the displacement of Gondwana—encompasses the entire area south of 4. The subordinated flux of retrograde appearance encompasses the entire width of 9 and 10, as well as of 8 and its extensions as far as the Himalayas. It is also of direct orientation but with lesser velocities than it displays in the main deep-seated portion of Gondwana, which underthrusts it, and it is only with respect to this main part that it should be considered as retrograde. The tendency of the Austro-Alpine flux to move beneath the Apennines, opposite the Ligurian promontory, also is outlined.

Figures 20A and 20B pertain more particularly to a time just preceding the end of the Jurassic, the epoch of the first important Andean deformations of the Tethys. With respect to the progression of the fronts in the Alpine-Carpathian loop from the Jurassic to the Quaternary, see by Emile Argand the memoir entitled "Plissements précurseurs et plissements tardifs des chaînes de montagnes," *Actes Soc. Helv. Sc. Nat.*, Vol. 101, Part II (1920), pp. 13–39.

Figure 21: The Mediterranean Frame after the Major Activity of the Great Tertiary Paroxysms and before the Great Disjunctions

Legend. 2: Horizontal projection of the thrusted frontal margin of Gondwana. P, Q: See Figures 20A and 20B. 9': Pennine involution (6' of Figure 26). 5, 7, 8, 9, 10, 11 and dotted area: See Figure 26. For more details on this epoch, see Figure 26 and its explanation.

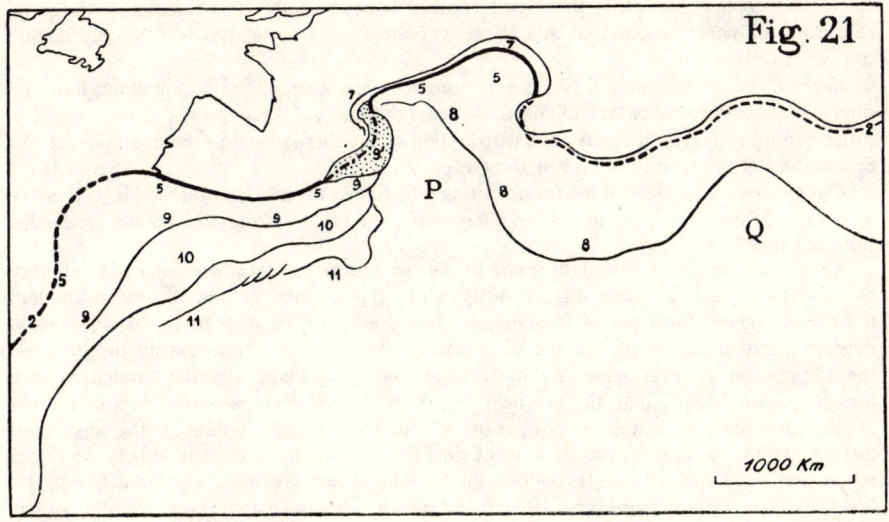

Figures 22 and 23. The Mediterranean Frame during the Great Distensions and Disjunctions

This powerful deformation, which has recreated at the expense of Gondwana and Eurasia—strongly engaged one upon the other—a deep Mediterranean, results from a drift of Europe toward the north. Its duration is sometimes designated in the text as the second phase—the great Cenozoic paroxysms being in this sense the first—or the phase of the great tearings. It is during this span of time that the Mediterranean frame acquires the essential features of today; all the deformations that this frame will subsequently undergo will be only details compared to that phase.

Legend. 1: Thinned sial. 2: Sial more thinned on the average, with or without holes of sima in places. Because of lack of space, the least thinned sial is not always shown along the entire margin of narrow objects. The distribution of the thinned sial in the main part of old Eurasia and of old Gondwana is not shown.

Figure 22 shows a state of the frame during which the powerful deformation is well on its way; Figure 23 corresponds to a subsequent state, the great deformation having reached a more advanced condition.

Short presentation of a general problem. Distensions consist of a lenticular cutting up of the sial (see page 135 and Figure 19B) while the surface of the distended domain is lowered. These facts are of fundamental importance to the diagnosis and chronologic determination of the phases of the deformation. At any given time during the process, the depression is maximum in the central, or axial, zone of the trough, where the distension began, and this vertical effect gradually decreases toward the margin of the distended domain. In the case of very advanced evolution, the distended domain includes a central, or axial, zone of sima (I) followed, toward the periphery, by zones of less and less distended sial: II, strongly distended and very depressed sial, grading into III, less distended and less depressed sial, beyond which extends, in continuity, the sial not yet reached by the distension in progress (IV). With the passing of time, zone I extends, as well as the entire area undergoing distension; zones II and III are gradually displaced toward a peripheral area, which III is conquering over IV, and all the corresponding bathymetric and altimetric conditions are displaced in the same direction. In the case of a less advanced evolution, zone I and even zone II may only be present; but this does not change the general character of the displacement of the vertical effects toward the periphery. The sea covers zones I and II; it may cover III entirely or only partially; in this second case the shoreline defines by default the margin of the distended area. Besides, the sea may also cover portions of IV; the shoreline in this case defines by excess the margin of the distension basin. Thus, in the second case. IIIa being the submerged part and IIIb the emerged part of III: IIIb is the location of an intense nonmarine clastic sedimentation, in general poor in fossils and fed by areas belonging to zone IV, which remained at a higher elevation (see the major nonmarine portion of the Aquitanian around the Mediterranean). *Any distension in progress that does not have a strictly intracontinental location determines a marine transgression* whose wedge records at any instant the increase of the distended area (see the rare patches of marine Aquitanian visible along the margins of the present Mediterranean and the broad Burdigalian transgression). This recording takes place, according to times and locations, by excess or by default. The bathymetric conditions pertaining to II and to III (or IIIa) being displaced toward the periphery, a distension in progress is expressed—in a vertical profile at a given location in the area it involves—by deposits whose succession and nature indicate a lowering

Illustrations 193

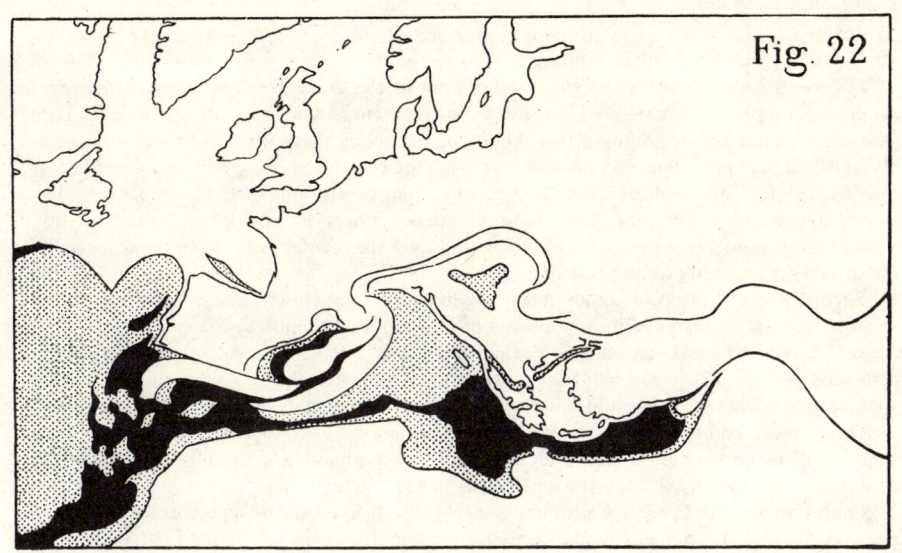

Fig. 22

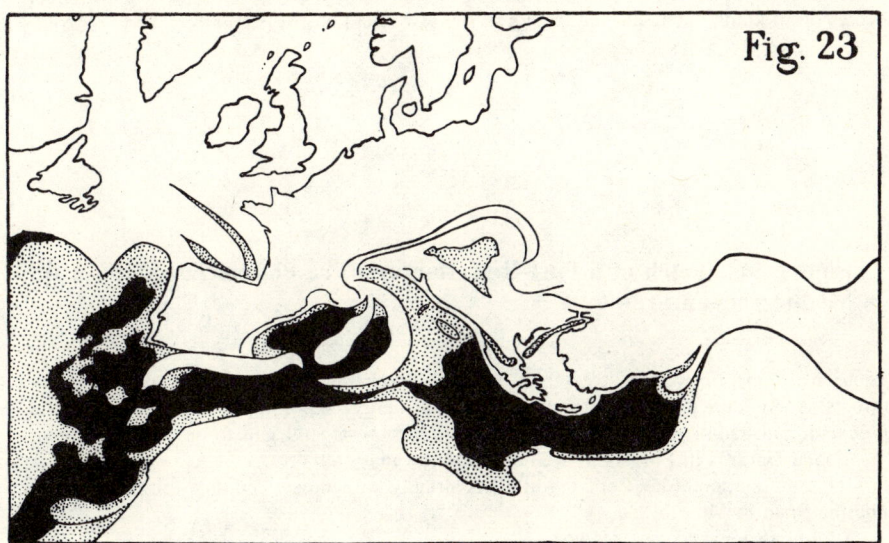

Fig. 23

1   1000 Km   2

(for example, the rather general deepening of the Mediterranean margin at the transition Burdigalian-Lower Helvetian, with its Schlier facies).

Diagnosis and chronology of the main phases of the great Mediterranean distensional process. First phase: symptoms of local or regional extent and of moderate character in Late Oligocene and at the beginning of the Aquitanian. Among these first centers of distension several shall become, in the second or third phase, the central, or axial, regions of great deep troughs; these centers will become deeper, will enlarge, will intersect, and will finally become integrated into complex basins of great extent. Others, like the trans-Aegean furrow, will not go beyond a more modest condition, and their narrowness will make them susceptible in certain segments to obturations by folds.

Second phase: Late Aquitanian times. Multiplication, extension, and enlargement of the serious injuries; appearance in new places of still moderate injuries. Figure 22 represents approximately the condition reached during this phase. The subsequent enlargement of the distended areas leads to the fact that most of the Upper Oligocene or Aquitanian marine wedges are beneath the present Mediterranean, the others being detectable in small number.

Third phase: end of Aquitanian or Early Burdigalian times. Generalization of the distensible and disjunctive stresses that reach their peak development. The essential features of the new planimetry are acquired at the end of this phase (Figure 23).

Fourth phase: Burdigalian. Continuation of the distensional and disjunctive stresses but in a somewhat less brutal manner. The complex areas of distension continue to encroach upon the margin of the dismembered fragments and upon the two main continental masses; hence the distribution and the amplitude of the Burdigalian transgression.

Fifth and final phase: Early Helvetian. The vertical effect of the distensions, whose average intensity has further decreased, remains apparent through the rather general replacement of a more shallow sedimentation regime by the Schlier. The maximum of the effects of all kinds, including the results of the previous phases, is reached.

Figure 24. **Sketch of a Post-Helvetian and Pre-Plaisancian State of the Mediterranean Frame**

There are moderate folding stresses in most of the chains and new advances of marginal thrusts. A minimum portion of the intracontinental energy—a portion that is distributed in the preexisting articulations and within a few new objects—is sufficient to develop the above-mentioned features in Eurasia as well as in Gondwana.

The continuation of the Betic cordillera is already looped between the Moroccan Meseta and the Spanish Meseta.

Atlantic distensions and disjunctions.

(From Late Helvetian included to Pontian included, the role played by tractions in the Mediterranean and its vicinity is intermittent, subordinated, and limited to a few regions. Folding behaviors—although moderate—predominate in general, here by means of narrow folds, there by means of broad basement warpings: the uplifting effects prevail. As is well known, this renewal reaches its maximum efficiency during the Pontian. But, the idea according to which the Mediterranean was by this process reduced to a narrow east-west trending channel is pure invention; the main troughs did not differ very much at that time from what

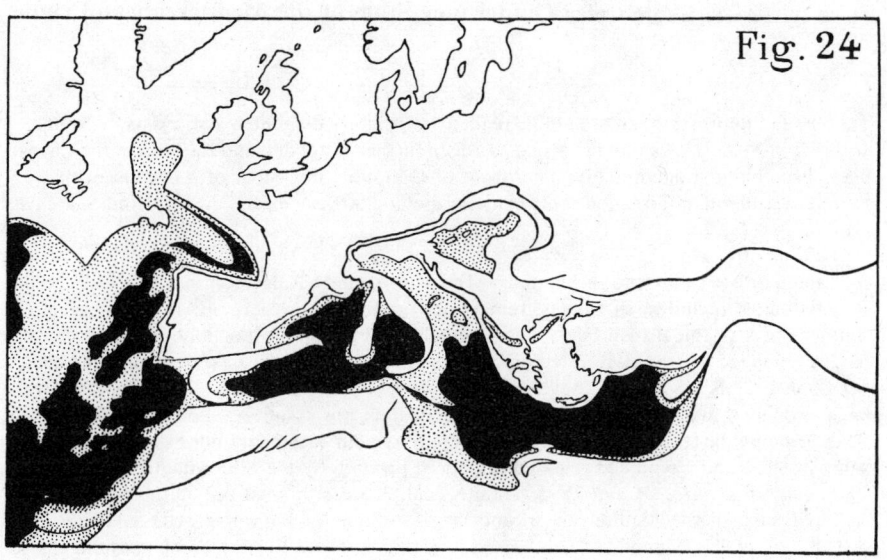

Fig. 24

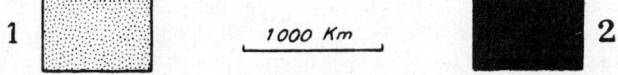

1   1000 Km   2

they had been in the Burdigalian or from what they are today. During the Plaisancian, new distensions and disjunctions occur, although of small importance if compared to those of the Early Neogene and recorded, along the margin that subsides, by the Plaisancian transgression.)

Figure 25. Sketch of a Quaternary State of the Mediterranean Frame

The Aegean depression, an area of distended and slightly disjointed sial, exists, as well as a *deep* Black Sea. The Pannonian space has been further distended and the loop of the Carpathians has enlarged accordingly. The Strait of Gibraltar, the effect of a distension that occurred in a meridian direction, between Europe and Africa, exists. A deep Red Sea exists also.

Folding stresses.

Atlantic distensions and disjunctions. The Mid-Atlantic Ridge continues to undergo differentiation; it includes, as fringes, remains of the basement folds of Armorica and of the southern part of the British Isles. These remains, which unquestionably contain some sial having belonged to original infratectonic lenses related to the large basement anticlines, have been appreciably stretched. They have more or less preserved their width, but they have been strongly thinned in the vertical sense; hence their low condition and the small degree of floating above the level of the sima. Remains of Pyrenean basement folds have undergone the same process and have been released in a stern position by the Mid-Atlantic Ridge.

Legend of Figures 24 and 25. 1: Thinned sial. 2: Less thinned sial on the average, with holes of sima in the Mediterranean and large surfaces of the latter in the Atlantic. The distribution of the thinned sial, in the main part of both old Europe and Gondwana, is not shown.

The following general comments are of particular application to Figures 6, 19B, and 22 to 27.

The distensions, which in most cases precede the disjunctions, lead to a more or less effective deformation of the two lips of sial undergoing separation. This generates a first series of alterations of the congruence. Besides, the author of these restitutions has found reasons to assume that the width of many segments of slope, rather steep immediately after the disjunctions, increased or increases up to a certain point through time. The plasticity of the sial leads to the fact that there is, whenever this material is considered over great surfaces, a critical slope above which the marginal sial of the continents is unstable and tends, through a slow gravity-induced flow—sometimes accompanied by local slumping with scarps and gliding surfaces—to return to the critical slope, whose value reaches only a small number of degrees for the upper surface of great masses. These phenomena generate a second series of alterations of the congruence.

The existence of these different alterations is sufficient to suppress the objections one would base upon the degree of approximation of certain congruences. In the preparation of the figures, I took into account these alterations in all the instances in which this seemed warranted.

Fig. 25

Figure 26. The Mediterranean Frame after the Major Activity of the Great Tertiary Paroxysms and before the Great Disjunctions

Approximate scale. 1:18,000,000.

This map shows, with more details, the essential parts of Figure 21. Africa overthrusts Eurasia. The overthrusting of Africa has reduced the Mediterranean to a network of epicontinental seas and of furrows devoid of any great depth.

Legend.

1: Gondwana.

2: Eurasia still welded to North America.

3, 3', 3": Northern margin of Africa. 3: Same margin, to which have been restituted the parts destroyed today by erosion. 3': Same margin, buried under other nappes (the Pennine involution of the northern and central Apennines) or hidden today by covers (Oltenia). 3": Approximate location of the continuation of that same margin, in horizontal projection.

4: External margin of the Pennine zone. 4': Internal margin of the same zone.

5: Austro-Alpine nappes (the upper nappes of the Betic cordillera, Balearic Islands, and Alps; the upper nappes of the Carpathians and Rhodope massif).

6: Pennine nappes in the few regions in which they are not covered by other nappes. The adjacent blank space, in which the dotting has not been drawn for graphic convenience, has the same significance. 6': Extensive mass of backward folds and of retrograde nappes belonging to the Pennine zone of the Alps, underthrusted by Africa and by the true Apennines which belong to that continent. This involution of Africa is due to the strong resistance of the Ligurian promontory of Eurasia and to the fact that the African margin was perhaps in that segment slightly lower than it was in the adjacent segments. But old Eurasia is not underthrusted by Gondwana in that region; just the opposite occurs. One can see the Pennine material, at both extremities of the involution, being refolded at the front, then beneath the Austro-Alpine nappes, but eventually regaining completely its normal position. An involution due to a similar and symmetrical behavior exists perhaps in the Balkanic peninsula, opposite the Getic promontory, but in the present state of knowledge nothing more can be said.

7: Margin of the Helvetic nappes and of the nappes of the Carpathian Flysch. 7': Autochthonous of the Balkan. 7": Scars of Mesozoic and other deposits (see explanation of Figure 27).

8: Dinarides.

9: True Apennines, that is, arisen from the slope of Africa.

10: Algerian Atlas and its extensions.

11: Saharan Atlas forming cover folds less pronounced than today.

12: Sheaves consisting of the left wings of the virgations of the second type by means of which the Saharan Atlas, the Algerian Atlas, the Apennines, and the Pennine involution terminated one after the other, following an echelon pattern against the western margin of the main part of the African promontory. As to what happened to these sheaves, see the explanation of Figure 27. No tectonic continuation of the Apennines into the Dinarides, but continuity of the stratigraphic material. Apennines and Dinarides end face to face in upper Italy by free extremities that are too weak to complete—with a small radius of curvature in plan—the difficult turning around the northwestern extremity of the most massive portion of the African promontory.

13: Granitic Corsica.

14: Esterel, which is together with the Mercantour and granitic Corsica the strongest intumescence of the Ligurian promontory, which continues eastward under the Alps.

15: Sardinia.

Illustrations 199

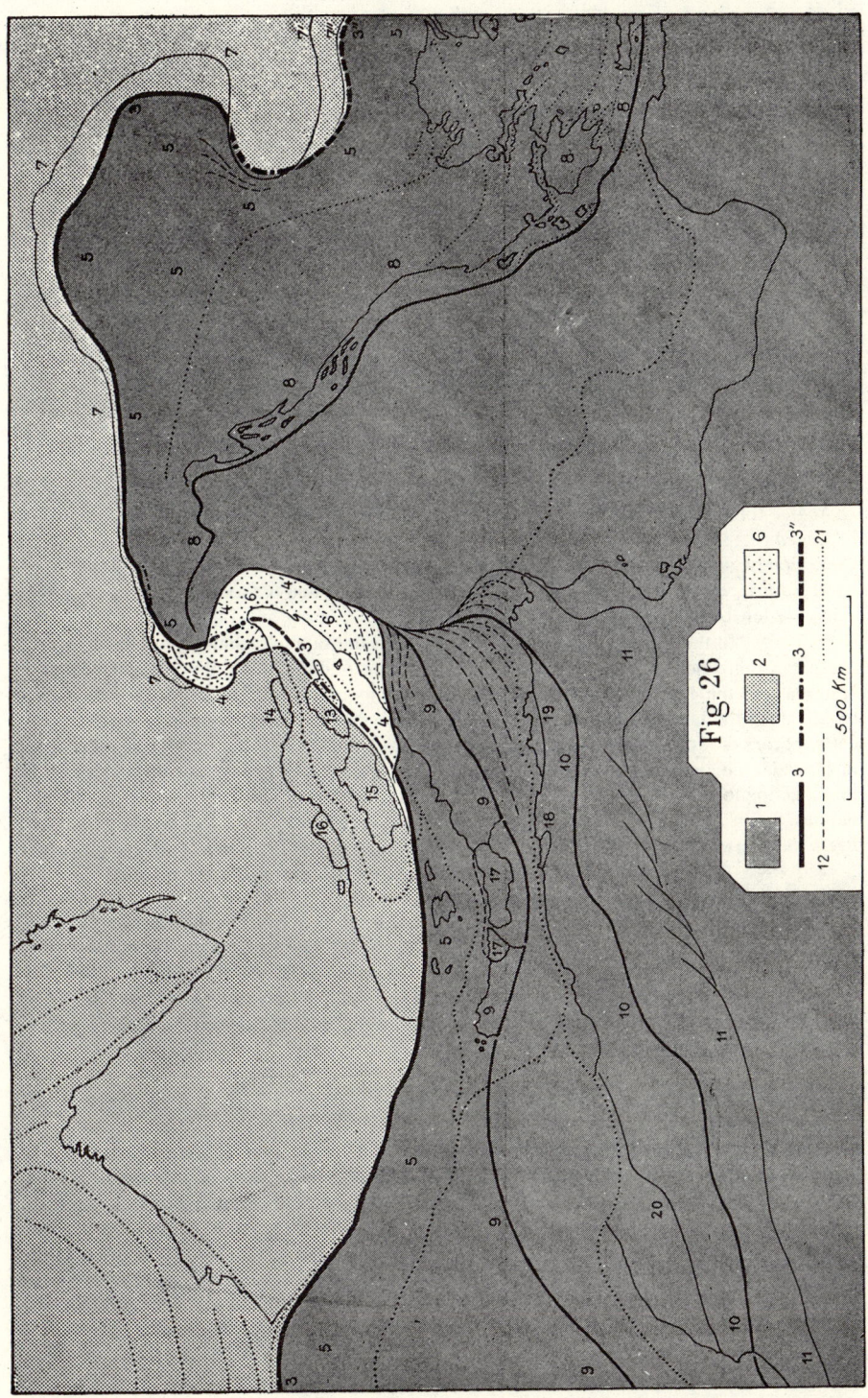

Fig. 26

16: Massif of Catalonia.

17: Massif of Calabria and of the Monti Peloritani, consisting of nappes of the Alpine cycle.

18: Massif of Great Kabylia.

19: Massif of Philippeville and of Bône.

20: Moroccan Meseta.

21: Lines along which the distensions or the disjunctions of the great phase will begin. Inside the Aegean area, I have shown as an exception, together with lines of distensions that began with the great phase (the trans-Aegean furrow, the furrow of Vardar-Morava) the lines corresponding to subsequent distensions and disjunctions of lesser importance. But the lines of future disjunction indicated along the Ionian Islands, as well as between Messinia and Crete, correspond to very important tearings of the great phase, those that have opened the deep basins of the Ionian Sea and of the eastern Mediterranean.

Figure 27. Present State of the Mediterranean Frame

Scale. 1:18,000,000.
Legend.

1: Gondwana.

2: Eurasia. The most thinned parts of the sial of Eurasia and of Gondwana, as well as the sima, are left blank.

3, 3′, 3″: Northern margin of thrusted Africa, with the same detailed legend as in Figure 26. 3‴: Connection that recalls the large disjunction undergone by this margin between Sardinia and the Balearic Islands. The horizontal projection of the margin of Africa (3″) trends along the northern boundaries of the Rhodope massif. This margin is emphasized by narrow scars of Mesozoic and other deposits that reach along strike the Black Sea at the Gulf of Iğneada. The line as shown in this figure—very approximate on a plan view—does not prejudge the location of the margin in the vertical dimension. It leaves to the Austro-Alpine nappes, that is, to Africa, the Rhodope massif, with the old ancient mass of Istranca, and the *Byzantine massif,* namely, the Paleozoic of both banks of the Bosporus, with Istanbul.

4: External margin of the Pennine zone. 4′: Internal margin of the same zone.

5: Austro-Alpine nappes.

6: Pennine nappes in the few regions in which they are not covered by other nappes. 6′: Pennine involution of Africa, disjointed or distended in the southwest and west between the Italian peninsula and the Corso-Sardinian massif. The disappearance of the involution to the north, under the Austro-Alpine nappes, is concealed by the Tertiary and Quaternary deposits of the Piedmont. The cover folds of the Neogene of the hills of Turin, which curve to the south toward Moncalieri, were formed much later.

7: Margin of the Helvetic nappes and of the nappes of the Carpathian Flysch. 7′: Autochthonous of the Balkan. 7″: Scars mentioned when discussing 3″. 7‴: Autochthonous of the Alps, including the Jura.

8: Dinarides. The disjunctions of the great phase have left with Africa a segment of the front of the Dinarides of Greece, which segment extended, as shown also in Figure 26, from the vicinity of Messinia to that of western Crete and of the island of Gávdhos. This segment, strongly thinned and consequently depressed with the help of isostasy, is disclosed by the bathymetry north of Cyrenaica, to which it is still attached by a distended and rather narrow peduncle.

9: Apennines. In the peninsula as in Sicily, the medium thickness line defines the minimum width attributable to the nappes, the margin of the latter being generally concealed under Neogene covers.

Illustrations 201

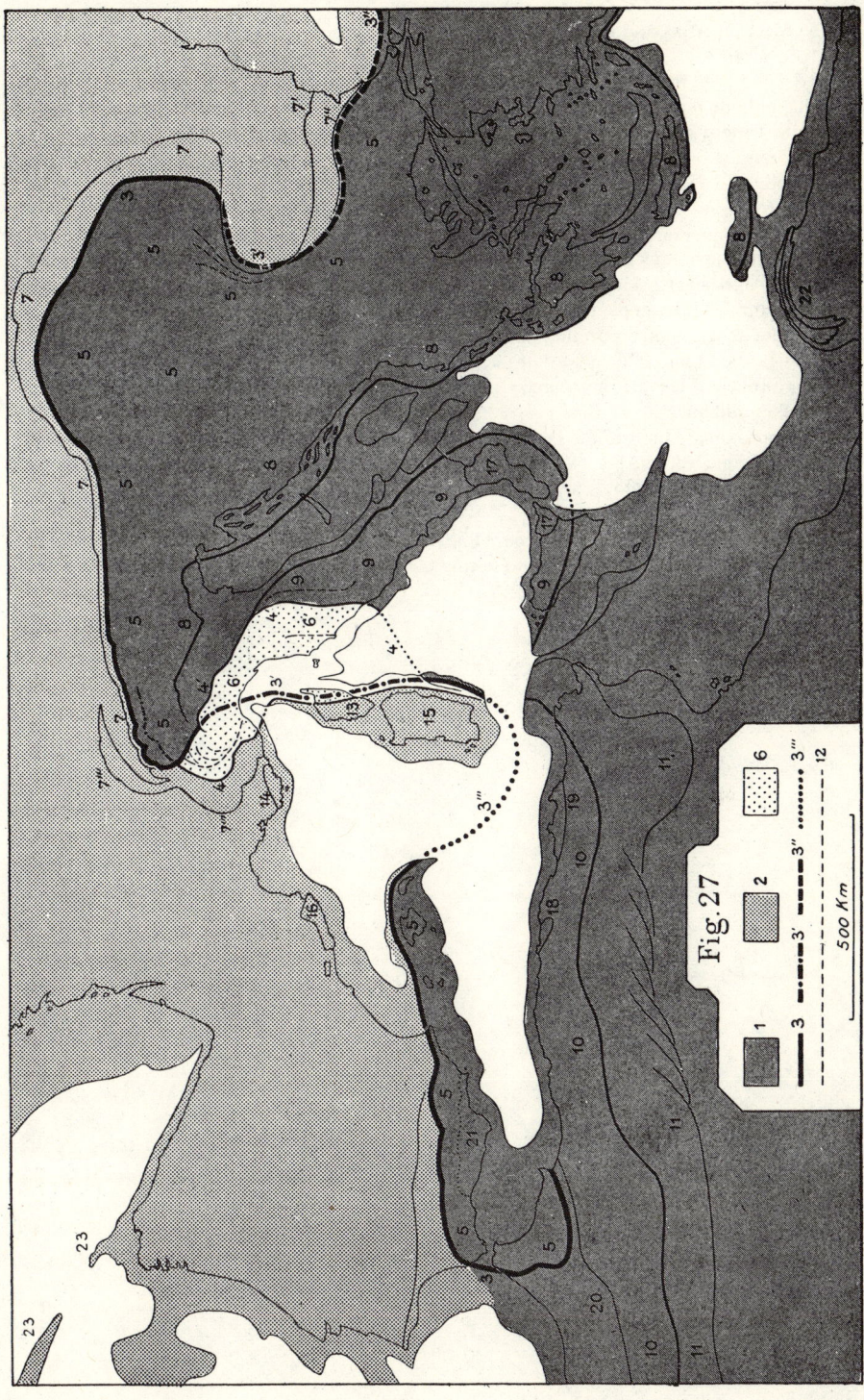

Fig. 27

10: Algerian Atlas and its extensions.

11: Saharan Atlas, consisting of cover folds, with the outline of a right wing of a virgation of the second type that extends at least from opposite Figuig to opposite Biskra and results from the oblique boarding of the Saharan basement by the flux coming from the northwest. This right wing was in the past symmetrical with the left wing mentioned below.

12: Virgation of the Banat and internal virgation of the western Alps due to overstepping of the Getic and Ligurian promontories by the flux, followed by its spreading in both corresponding reentrants. Meridian alignments of central Italy, oblique with respect to the front of the chain: they are deformed remains of the virgation of the second type displayed, on the left wing, by the Apennines and their Pennine involution. The remains of similar virgations by which the Algerian Atlas and the Saharan Atlas ended at their left wing have been thinned, together with the African promontory, then engulfed below the Apennines, which were drifting to the east while incorporating this thin sial.

13 to 19: See Figure 26: note that the positions of these objects have changed.

20: Moroccan Meseta, confronted since the Helvetian by the Rif, looped continuation of the Betic cordillera.

21: Approximate limit of the Pennine window of the Sierra Nevada, in continuation with the Pennine zone of the Alps. With the exception of the massif of Calabria and of the Monti Peloritani, the other windows are not shown.

22: Cover folds of Cyrenaica.

23: Stretched remains of the basement folds of Cantabria and of the Pyrenees, with traces of the tearing-off peduncle that formed during the traction generated by North America and later by the Mid-Atlantic Ridge.

Notes
and
Analytical Index

Notes

La Tectonique de l'Asie being the text of an address, Argand kept his references to a minimum, giving generally just the author's name without date or source. All references have been checked for accuracy and are presented in full. Argand's notes are identified by (au), standing for author. All other notes are additions by Albert V. Carozzi.

1. E. Suess, *Das Antlitz der Erde*, 3 vols. in 4 (Prague: F. Tempsky, 1883-1909); *The Face of the Earth,* translated by Hertha B. C. Sollas under the supervision of W. J. Sollas, 5 vols. (Oxford: Clarendon Press, 1904-1924).
2. M. Bertrand, "La Chaîne des Alpes et la formation du continent européen," *Bull. Soc. Géol. France,* Ser. 3, Vol. 15 (1887), pp. 423-447.
3. For reference purposes, the page numbers of Argand's original text are retained in the English translation and are enclosed in parentheses. Each page number refers to the material following it.
4. E. Argand, "Sur l'arc des Alpes occidentales," *Eclogae Geol. Helv.*, Vol. 14, No.1 (1916), pp. 145-191.
5. E. Argand, "L'Exploration géologique des Alpes Pennines centrales," *Bull. Soc. Vaud. Sc. Nat.*, Vol. 45, No. 166 (1909), pp. 217-276; also *Bull. Lab. Géogr. Phys. Univ. Lausanne,* No. 14 (1909), pp. 1-64; idem, "Les Nappes de recouvrement des Alpes occidentales et les territoires environnants—Essai de carte structurale 1:500,000," *Mat. carte géol. Suisse,* N. S., Livr. 27 (Berne, 1911) (carte spéciale No. 64); idem, "Les Nappes de recouvrement des Alpes Pennines et leurs prolongements structuraux," *Mat. carte géol. Suisse,* N. S., Livr. 31 (Berne, 1911), pp. 1-26.
6. For a typical example of modern graphic expression in the field of tectonic maps, see N. Schatsky et al., "Carte tectonique internationale de l'Europe, 1962," Congrès Géologique International, Commission de la Carte Géologique du Monde, Sous-Commission de la Carte Tectonique du Monde (Moscou: Académie des Sciences de l'URSS, 1964).
7. (au) E. de Margerie, "The Geological Map of the World," *Compte-Rendu XII [1913] session Congrès géol. international* (Ottawa, 1914), pp. 173-187; idem, "Une Nouvelle carte géologique du monde?" *Annales de Géographie,* Vol. 31, No. 170 (1922), pp. 109-131.
8. (au) In some of the sections below, I have developed to a certain extent interpretations that the tectonic map of Eurasia, presented during this talk and posted for the duration of the congress, suggests or clearly displays.
9. The term "crystalline" (*cristallin, terrain cristallin*) is used by Alpine geologists for the general designation of the association of metamorphic and igneous rocks regardless of their tectonic behavior.
10. The term *bâti,* which means essentially "constructed object," is frequently used by Argand for designating materials folded and indurated by one or several successive orogenies and against which later structures will mold

themselves. These objects may play the role of obstacles, but they are not necessarily synonymous with basement; hence the translation of "frame" used here.

11. (au) This term having been considered in different manners within a single work, some precision is required. The Yarkand arc is the segment of old formations, concave to the northeast, that from approximately the 74th to the 82nd meridian (transverse alignment Keriya-Kyzyi Daban) borders the plains of Chinese Turkestan and buttresses to the southwest and south the Alpine foldings of Carboniferous, Permian, and Mesozoic formations of the Tethys that follow the northeastern margins of the Mustag-ata and of the Karakorum. In this sense, the Yarkand arc is identical to the western Kunlun of several authors.

12. E. Argand, "Plissements précurseurs et plissements tardifs des chaînes de montagnes (Discours d'ouverture)," *Actes Soc. Helv. Sc. Nat.*, Vol. 101, Part II (Neuchâtel, 1920), pp. 13–39.

13. (au) Ch. Jacob, "Etudes géologiques dans le Nord-Annam et le Tonkin," *Bull. Serv. Géol. Indochine,* Vol. X, Fasc. 1 (1921), pp. 1–204.

14. The term *rebroussement* has been translated by "backward deflection," which is preferable to "syntaxis" used by E. Suess, *The Face of the Earth* (see note 1 above), for the German *Schaarung* (Vol. I, p. 422) and also by W. H. Bucher, *The Deformation of the Earth's Crust* (Princeton: Princeton University Press, 1933; reprint ed., New York: Hafner Publishing Co., 1957, pp. 80–81.

15. E. Argand, "Les Nappes de recouvrement des Alpes Pennines et leurs prolongements structuraux," *Mat. carte géol. Suisse,* N. S., Livr. 31 (Berne, 1911), pp. 1–26.

16. E. Argand, "Sur l'arc des Alpes occidentales," *Eclogae Geol. Helv.*, Vol. 14, No. 1 (1916), pp. 145–191.

17. E. Suess, *Das Antlitz der Erde* (see note 1 above).

18. Serindia is a large east-west trending depression of central Asia between the Pamirs and the Pacific watershed and bordered to the south by the Kunlun and to the north by the Tien Shan. It corresponds essentially to the Takla Makan Desert and the Tarim basin. In recent tectonic maps, it is called the Tarim Stable Block.

19. (au) What I designate by this name is not only the territory of the Ordos; it is the entire huge promontory that forms the southwestern region of the Sinian massif and spreads out extensively inside the great loop of the Huang Ho, encroaching upon adjacent areas.

20. The French word is *Anthracolithique,* which in spite of its etymology has been used, after E. Haug, for the Permo-Carboniferous System whenever further subdivision is not possible.

21. B. Willis, "The Mechanics of Appalachian Structure," *13th Annual Report of the Director of the U.S. Geological Survey* (Washington, D.C.: U.S. Government Printing Office, 1891–1892), published (1894), Pt. II, pp. 211–281.

22. B. Willis, *Research in China* . . . , 3 vols. in 4 (scientific results of the Carnegie Expedition of China, 1903–1904, Carnegie Institution of Washington, Pub. No. 54 (1907–1913).

23. See note 22 above.
24. See note 21 above.
25. (au) W. Bowie, "Investigations of Gravity and Isostasy," *U.S. Coast and Geodetic Survey, Special Publication No. 40* (Washington, D.C., 1917), map 13.
26. A. Wegener, *Die Entstehung der Kontinente und Ozeane* (Braunschweig: F. Vieweg und Sohn, 1915), Sammlung Vieweg No. 23, 94 pp.; there are numerous subsequent editions; idem, *The Origin of Continents and Oceans*, translated from the third German edition by J. G. A. Sherl (London: Methuen and Co., 1924), 212 pp.
27. Argand introduced here the term "mobilism" to characterize the concept of continental drift as opposed to the traditional view of "fixism," which assumes that the continental blocks never changed position through geologic time.
28. The French terms are *jeux de proue* and *jeux de poupe*, respectively translated as "bow stresses" and "stern stresses." Argand compares here the effects of drifting sial rafts to the eddies and other currents created in water by ships in motion.
29. A. Heim, "Die Schwereabweichungen der Schweiz in ihrem Verhältniss zum geologischen Bau," *Vierteljahrsschr. Naturf. Gesell. Zürich,* Vol. 61 (1916), pp. 93–102.
30. A. Wegener, *The Origin of Continents and Oceans* (see note 26 above), p. 159, fig. 31.
31. E. Argand, "Les Nappes de recouvrement des Alpes Pennines et leurs prolongements structuraux," *Mat. carte géol. Suisse,* N. S., Livr. 31 (Berne, 1911), pp. 1–26.
32. E. C. Abendanon, *Die Grossfalten der Erdrinde* (Leiden: Verlagsbuchhandlung vormals E. J. Brill, 1914), 183 pp.
33. P. Termier, "La Synthèse géologique des Alpes" (1906) and "Les Problèmes de la géologie tectonique dans la Méditerranée occidentale" (1911), in *A la gloire de la Terre* (Paris: Desclée de Brouwer et Cie., 1916), pp. 45–114.
34. L. Gentil, M. Lugeon, and L. Joleaud, "Les Nappes de charriage du bassin du Sebou (Maroc occidental)," *C. R. Somm. Soc. Géol. France,* Vol. 18 (1918), pp. 115–117.
35. (au) Sir John Murray and J. Hjort, *The Depths of the Ocean* (London: Macmillan and Co., 1912), map III.
36. (au) If one rejects, in contradiction to all appearances, the idea of a difference in nature between oceanic depths and continental masses, the concept of frame or of framed mass remains nevertheless related, on this upper scale, to the depths under discussion. This is true also if one rejects mobilism. All things considered, no known fact compels us to deny that these depths possess an aptitude for folding.

Assuming mobilism to be true, nothing allows us to say at present whether the frame in the upper indurated parts of the sima can undergo folding or not. The question remains open. In the case of a positive answer, the point would be to know how long such folds could maintain themselves. Since this would depend on the consistency of that medium and of that of the underlying sima, that is, on factors that escape almost completely present understanding, it is better to drop the subject.

(Argand added this note during the printing of *La Tectonique de l'Asie*.)

37. This paragraph is an excellent expression of Argand's philosophy, namely, his belief in the gradual development of science, in the lack of the so-called scientific revolutions, and, finally, in an orderly cosmos expressing perhaps some divine organization.

Analytical Index

Introductory comments: the orogenic cycles   1
   I. Of the art of interpreting structural facts. Static tectonics and dynamic tectonics. Vision and theories. No synthesis of the movement without the vision of a continuum in three dimensions undergoing deformation. That the tectonics in movement, the deformation in progress, regulates the major events of the stratigraphic order and of the morphologic order. Of the three aspects of any deformation. Of the vertical aspect: the axial behavior of folds and the role of obstacles. Of the insignificance of the greatest vertical faults with respect to the dimensions of the deformed volume. Predominance of the horizontal aspects. Of resorting to epeirogenic movements and to radial dislocations. Fragility of these hypotheses and why they should be questioned. Predominance of the deformation in volume over fracturing. Of tabular lands. Of real transverse folds and of transverse traces. Of backward foldings. Of geosynclines whose history encompasses more than one orogenic cycle (polycyclic geosynclines). Of the unequal sensitivity of the recorders of movement. Of the vision of the total deforming movement   2
  II. The tectonic map of Eurasia on a scale of 1:8,000,000   10
 III. Short description of Precambrian Asia and of Caledonian Asia. The Siberian massif and its surroundings. Limit toward the arc of Taymyr. Limit toward the arc of Verkhoyansk, fragment of a Periarctic Alpine chain. Of the Sinian massif. Marginal foldings of the amphitheater of Irkutsk. Of the problem of the Siberian-Mongolian crests. Pre-Devonian fragments included in the Kunlun. Caledonian fragments in Burma. Influence of the Caledonian orogeny on the old massifs and on the Himalayan geosyncline. Of the Caledonian folding in Asia and outside Asia. The Proto-Atlantic as Caledonian geosyncline. Of the depression between the Russian platform and the Siberian massif during the Caledonian cycle   12
 IV. Short description of Hercynian Asia. Urals, massif of the Kirgiz steppes, Russian Altai, Tarbagatai, Dzungarian Alatau. Tien Shan, Kunlun, China. The common Hercynian trunk of central Asia. Its division into branches to the west. Of the delimitation

of the pre-Hercynian massifs responsible for this division. Hercynian folds of the crest. Influence of the Hercynian orogeny on the old massifs and on the Himalayan geosyncline 22

V. General features of the Alpine cycle. Of the Andean subcycle. Short description of the Andean foldings, of the Laramide foldings, and of the Alpine foldings properly speaking in the chains of western America. Two main generations of new folds: one of Andean age, the other of properly Alpine age. Two generations of basement folds: one of Laramide age with mainly Precambrian indurated material, the other of properly Alpine age with mainly Andean indurated material. Distribution of these objects; sketch of the movement. Synergy of the Andean movements of America and of Eurasia. Of movements belonging to the Alpine cycle but whose upper age limit cannot be precisely established in the present state of knowledge: Malacca peninsula, Indochina, China, Korea (with a glance at ante-Alpine foldings), Manchuria, eastern Siberia. Andean movements: Japan, Himalayas, northern Siberia. Pre-Lutetian movements: synergy with the Laramide foldings; case of the Alps 24

VI. Of the behavior of the Tethys and its continental *rear zones* at the times of compression. Power and resistance. Outline of a distribution of the energy between folding objects of different categories. Intracontinental energy. Energy of compression. Energy of new chains. Energy restituted to continental margins 30

VII. Of the behavior of the chains of the Tethys: general description. Particular cases: Himalaya, Tibet, southern Iranian arc, arc of Mascate. Cover foldings of Indo-Africa. Asiatic extension of the northern wing, displaced to the north, of the double Mediterranean chain 34

VIII. Of virgations. Virgations of first type. Virgations of second type. Definitions. Shapes. Outline of a deformation regime. Particular cases. The Iranian virgation. The Afghan virgation. Interpretation of the local arcs that complicate the eastern wing of the southern Iranian arc, and a glance at the major axial behaviors of that arc. How the virgations allow us to determine the general direction of flow of the plastic fluxes. Virgations of cover folds. Virgations of recumbent folds. Virgations or cordilleras. Virgations of basement folds 37

IX. Of the deformation of old Eurasia, north of the Tethys, during the Alpine cycle: general description. The northern margins of the Asian Tethys from the Permian to the eve of the Tertiary paroxysms 43

X. Of *cover foldings:* deformation regime; Asian examples and others; comparisons. Of *basement foldings:* definition; shapes.

Of *ordered* chains. Features common to basement folds and to ordinary folds. The great axial behaviors of the chains of western America: interpretation. Explanation of axial behaviors, in general, by the combination of tangential effort and of resistances facing it. Difficulties of the epeirogenic theory. Culmination and depression of axes, caught in their very movement, are vertical aspects of folding in progress. Of *voussoirs of basement folds*. That over the entire Earth, there is not a single fracture whose radial origin is above all contestation. Nappes with clean-cut thrusts, derived from basement folds. Swellings that precede these thrusts. *Reactivation* of old frames by basement folds. Reactivated forelands. Energetic predominance of basement folding over that of new chains. These chains, and particularly those issued from geosynclines, become features of detail subordinated to basement folding. Basement folding works upon the entire mass of continents: it is the specific reaction of continental masses to the tangential effort and the main expression of folding on this   45

XI. The duel between Indo-Africa and Eurasia: characteristics and general descriptions. Short abstract of Paleozoic events: how old Eurasia was built. The duel during the Alpine cycle. Specialization of the initial segments through times into shorter segments dedicated to particular tasks. The more general features of the distribution of vertical effects in Eurasia during the Mesozoic and the Nummulitic, before the Tertiary paroxysms. Differences presented in that respect by old Europe, old Asia or Angara Land, and the intermediate segment. The axial behavior of the embryonic basement folding accounts for these differences without epeirogenesis. This approach locates the differences in the whole deformation whose unity is made apparent. Of the embryonic basement folds of the southern margin of Angara Land: relationship with the movements of India during the Early Mesozoic. Of the chronology of the Tertiary paroxysms in Asia. Of the Tibetan intumescence: structure in double chain; two wings with opposite overturning. Median location of the crushed Tethys. Two margins with basement folds: the Himalayan zone and the Kunlun. Analogy with the Mediterranean system. Extension of the folds over the course of time   53

XII. Of flux segments. Influence of obstacles and of channels on tectonic segmentation. Visual comparison with the flow of a river hindered by its banks, by the piers of a bridge, and by irregularly distributed obstacles. Picture of axial behaviors. Of flux transverse alignments. Of the concept of axial line. Linear schematism and spatial vision   57

XIII. *The segment of central Asia:* delimitation; general features of the deformation regime proper to this segment. Of the *Turanian segment*. On the condition of East Asia with respect to the segment of central Asia 59

XIV. Reactivation of the old frames of Eurasia during the Alpine cycle: role of the very old massifs and of the ex-geosynclines. Division of the segment of central Asia into two shorter segments whose behavior becomes more pronounced with the drawing together of India and of Angara Land: the *Indo-Siberian segment* and the *Indo-Mongolian segment*. Role of the Siberian massif in this event. General features of the flow regimes proper to the two new segments. The great axial behaviors: illustrations drawn from the Russian Altai, the crests, the Tien Shan, the Gobi, and the Tibetan intumescence. *Serindia,* the *Indo/Serindian space,* and the *Serindo-Siberian space* 61

XV. Serindia: delimitation. Basement folds that enclose the Serindian space. The problem of the deep-seated part of Serindia: small number of observable data. That this problem can be solved only by understanding the behavior of the foldings that surround Serindia both nearby and far away. The southeastern side of Serindia and the virgation of the middle Kunlun. The Nan Shan, the Altyn Tagh, and their rear chains build the major part of that virgation. The deep flux that carries the Kunlun flows to the northeast. The southwestern side of Serindia: the arc of Yarkand and the new Alpine elements located upstream. Overturning to the northeast toward Serindia 63

XVI. The northern periphery of Serindia. The distribution of the Hercynian deformation in the Tien Shan implies the existence of an obstacle located beneath Serindia. The Tien Shan belongs to a larger system, the *Turanian virgation*. It is the major element of the left wing of that virgation. The central segment of the virgation covers western Siberia and the Turan. General characteristics of the folding behaviors in that segment. The central transverse alignment of the Turanian virgation. The demand for matter upstream of the Turanian flux: the massif of the Kirgiz steppes. The influence of the margins of the Turanian corridor: general features 66

XVII. Two Turanian virgations: one Hercynian of new folds, the other Alpine of basement folds. The pre-Hercynian frame is common to both virgations. The right margin of the Turanian corridor: Russian platform, Podolian massif, and Arabian massif. The left margin of the Turanian corridor: Siberian massif, massif of the crests, Serindian and Indian massifs. The influence of the massif of the crests: Hercynian plan of the Russian Altai; adaptation to the salients of the Alatau of Kuznetsk and of the Mongolian Altai as well as to the intermediate reentrant. Alpine deformation of the Russian Altai: differences and analogies with respect to the Alpine deformations of the Kirgiz massif and the Tien Shan and its rear chains. Explanation of these facts by the conformation of the margins and the variety of conditions imposed on the flow of flux. Increasing effect of basement folding while nearing Serindia. This fact confirms the concept of a resistant Serindian massif. The Tien Shan swollen into superb basement folds results from the confrontation of Serindia by the flux. Resolution of basement folds into clean-cut thrusts that override Serindia. The behavior of basement folds is more constrained in the Tien Shan, freer in the rear chains (Tarbagatai, Saur, Dzungarian Alatau). The differences are explained by the distance to the Serindian obstacle, the analogies by the conditions common to the entire flux. The northern branches of the Tien Shan: the basement fold regime is intermediate between that of the rear chains and that which characterizes the main part of the Tien Shan. Dzungaria. The Bogdo-Ula and the arc of Bar Köl. Long-distance influence of the promontory of Irkutsk. The Turanian flux and the Serindo-Altaic flux, elements of the total flux. Their interaction, their distribution by regions, their effects. How the massif of the crests and the Serindian massif have influenced the distribution of these deformations. General plan of the virgation in the Tien Shan. The marginal thrusts of the Tien Shan; their axial behaviors. Fractures and longitudinal displacements. The stellate pattern of the Tien Shan 69

XVIII. Evaluation of the Serindian question. That an ancient Serindian massif exists which is the buttress common to the Tien Shan and the Kunlun pushed in opposite directions. The folding behaviors of the Indo-Serindian space and those of the Serindo-Siberian space have been dissymmetrical during the ante-Alpine cycles. The fan-shaped symmetrical structure visible in both spaces has been produced by the Alpine cycle. The narrowing of the Indo-Serindian space has been so important that the deformation regime has undergone a qualitative modification from it. Hence the necessity of assuming a very important horizontal displacement of India during the Alpine cycle. Of previous conditions: a quick glance at the Paleozoic Tethys. Of some differences between the Indo-Serindian fan and the Serindo-Siberian fan. The Alpine deformation of the crests, by basement folds, is complicated by important clean-cut thrusts oriented toward the interior of the Siberian amphitheater. Discussion of a few facts pertaining to the great horizontal movements reported in that region. Concept of a Siberian massif, heterogeneous at the crests and forming a buttress. These Alpine thrusts of Siberia are symmetrical to those of the Tien Shan, hence confirmation of the compression that took place between the Serindian massif and the Siberian massif 75

XIX. How the opposed fluxes of the Tien Shan and of the Kunlun meet east of Serindia in the Sino-Serindian channel. The flux of the Kunlun, flowing to the northeast, becomes predominant in the Nan Shan. Role of the Ordos, southwestern spur of the Sinian massif. The great compression of the Galbyn-Gobi. The virgation of the Ala Shan 80

XX. *The Sino-Siberian space*. Whether the Hercynian geosyncline of the Tien Shan extended, in a continuous manner, in that space to the Sea of Okhotsk. Of basement folding in the Sinian massif and in the Sino-Siberian space. Sikhota-Alin, northern Manchuria, Amurian lands. Hypothetical relationships of the Serindian, Sinian, and Siberian massifs in very ancient times. The boundaries of the very old massifs are incompletely known, but the reactivations by Alpine basement folds allow us to guess something about them. Concept of *nuclear masses*. How the Sinian massif reacted to the Alpine basement folding; differences of behavior between the margins of the massif and its central region. Tentative interpretation of the Sino-Siberian space by means of the concept of polycyclic geosyncline and by that of nuclear mass. Of some particularities of ante-Alpine folds in that space and in adjacent regions 82

XXI. General glance at the basement folds of the segment of central Asia. What they reveal of the behavior of basement folds in general. For great objects, the average plasticity decreases and the average degree of induration increases with increasing age of the frames. Discussion of this result: main factors of an explanation 86

XXII. The Urals, an element of the right wing of the Turanian virgation. Why the deformation of this wing is less intense and less rapid than that of the left wing. Deficiency and delay of the Hercynian deformations of the Urals with respect to those of the Tien Shan. Deficiency of the Alpine deformations of the Urals; incipient character of basement folding. Hercynian segmentation and Alpine segmentation of the Urals: fundamental role of the peculiarities of the eastern margin of the Russian platform. Timan and Novaya Zemlya. Mugodzhary and Tien Shan: the connection between Alpine basement folds is very certain but does not prejudge the precise location of the connection between Hercynian folds across the Turanian segment 88

XXIII. The Geguli-Ergheni arc. The Alpine folded elements of the steppes north of the Caspian Sea. Of the Russian platform. The essential behavior of the Caucasus: it is the Arabian massif which generated this basement fold and localized its culmination. Similar behaviors of Lusatia opposite the Precambrian massif of Bohemia   90
XXIV. The basement folds of Europe   92
XXV. New generalities on basement folding. Of the concept of plastic flux. Of the concept of undercurrent. Basement folding and stratigraphy. Basement folding and morphology. The *vertical effects* of basement folding and the currently accepted ideas on *vertical movements*. Discussion. The deformation in volume represents the real and the concrete; the idea of linear displacement is an analytical artifice except in the particular case of very small objects. That a large tectonic object cannot be displaced without being deformed. Of the hypothesis of eustatic movements. Of the deformation of extensive massifs of very large radii of curvature: domes, shields, and platforms. That these objects are or have been basement folds or complexes of basement folds. That the great shields, of the type of Fennoscandia or Laurentia, are essentially *basement brachyanticlines*. Synergy between the deformation that generates these large basement folds and that which molds the ordered basement folds facing them. The rule of axial behavior shows the reality of this synergy, of which many particular cases can be visualized. Tectonic deformation, isostasy, and gravity anomalies. Of the multiple interactions of these factors. Principal or essential behaviors and secondary or accessory behaviors. Importance of this distinction for the interpretation of the facts belonging to the three orders and for their synthesis. Criteria. Of the secondary character of isostatic behaviors regulated by external actions such as overloading by sedimentation or glaciation and unloading by erosion or deglaciation. That basement folding may be considered as a cover folding developed on an immense scale. Of *infratectonic* structures, result of the deformation at great depths. Necessity of hypotheses concerning the infratectonic symptoms of basement folding. Hypothesis of *sial* and of *sima*. The *lenses* of deep-seated sial are formed under great basement anticlines as well as under new chains, hence strong negative anomalies are generated: this is one of the essential infratectonic behaviors. Other essential behaviors under distended areas: thinning of the sial and rising of the sima. That secondary behaviors can lead to the reworking or the destruction of essential infratectonic objects. That these modifications, however, are far from predominating as a whole.

Of the interpretation of isoanomalies of gravity. That it is expedient to interpret the anomalies in the light of the visible tectonics. Dangers of the reverse process. Discussion of the Bouguer anomalies in the United States. Eastern Africa and Red Sea. Of extra-Alpine Europe. Of central Asia. Of the Himalayas and the Indo-Gangetic foredeep. In the zones of basement folds and of new chains, as a whole, the essential behaviors largely predominate over the secondary effects. Importance of axial behaviors in that respect. That the horizontal aspects largely predominate in the general deformation. Incapacity of linear schemes to account for any deformation whatsoever. The existence of axial behaviors, their reduction to a tangential behavior, and the fact that basement folding affects the entire mass of the continents prevent any reliable diagnosis of original vertical linear movements. The concepts of epeirogenic movements and of radial dislocation deprived of any reliable foundation. The explanation of basement folding requires, on the upper scale, a deformation resulting from essentially horizontal stresses encompassing all the continental masses and their adjacent parts 102

XXVI. Of the continuum and discontinuum in tectonics. The continuum is the form of the synthetic vision of movement; the discontinuum satisfies the requirements of analysis. Necessity and dangers of the imagery of a continuum undergoing deformation. That this imagery should be precisely checked by rational criticism. That it can express, by a single synthetic and condensed view, the summation of a multitude of small displacements developed on lower scales. East Asia: delimitation. The great protuberance of Southeast Asia, from Assam to Taiwan through Indonesia and the Philippines; Indochina. Of several analogies of behavior between this protuberance and the Iranian-Turanian segment. Southernmost China. The Chinling Shan. The recent foldings of Kansu and of western Shensi. That the festoons present as does all East Asia, but with a particular emphasis, the immense problem of the nature of oceanic depths and that of the fixism or mobilism of continents. Impossibility of avoiding the question, unless one restricts oneself, for large portions of East Asia, to an analytical catalogue of known facts 118

XXVII. Concrete tectonics and theories of the Earth. Two attitudes of the mind in front of the problem of continents: *fixism* and *mobilism*. Of several theories that imply fixism. Concept of the total deforming movement according to classical fixism. This concept, if assumed to be right, should be enlarged in order to encompass basement folding, a reality independent of any theory. The theory of contraction and basement folding. Outline of an extension of the classical approach; objections. Mitigated fixism; limited mobility of continents. Of some tectonic consequences of the hypothesis of a veneer of thin sial extending over the oceanic sima. Of the mobilistic theory, or theory of great continental drift   124

XXVIII. Soundings. The problem of the Pacific Ocean. Of some of the more general conditions that prevail in the suboceanic depths and, particularly, beneath the Great Ocean. That this problem can be approached indirectly, through the study of the variations of tonnage of the Circumpacific chains compared to the variations of the same factor in the chains arisen from the Tethys. Statistical criteria. Reactivated tonnage, new tonnage, total tonnage, normative tonnage, or tonnage per unit length of chain, energetic quota of a segment of chain: definitions, criteria, inferences. Of some significant facts. The fluctuations of the normative tonnage are more moderate, less subject to extremes in the Circumpacific belt than in the half-belt of the Tethys. Interpretation: the bottom of the Pacific Ocean consists, on the average, of a medium more homogeneous and more yielding than that of the upper parts of continental frames. No collapsed Pacific continent. No Circumpacific geosyncline. That there is a noticeable difference in behavior between Circumpacific chains and those issued from the Tethys. Of *marginal* chains, products of the folding deformation of continental margins, in particular of their slopes. Initial monoclinal arrangement of the marginal chains. Analogies and differences of behavior between these chains and those originating from geosynclines. Two kinds of new chains: geosynclinal chains and marginal chains. Outline of the distribution of energy in the marginal chains and in the portions of the continental margin facing them. Control of the folds in these chains. Agreement between the preceding interpretations and mobilism, and the new facts that support them. Other significant facts. The normative tonnage of the Circumpacific chains in East Asia is less than that of North America. Agreement with the distribution of *bow stresses* and *stern stresses* requested by mobilism. Faced with these new facts, fixist contractionism can call upon randomness only. The normative tonnage of Andean age. The mobilistic theory and basement folding. The great effects of traction as negative aspects of basement folding. Mobilism and the concept of geosyncline. The *drift over lenses* and axial behaviors. Axial behaviors on the scale of basement folds and their analogues on the continental scale. Of the complication of plastic fluxes. Axial behaviors and the great disjunctions   128

XXIX. Glances. The Bouguer anomalies in the United States and the drift over lenses. The Caledonian Proto-Atlantic and the present-day Atlantic. The Mediterranean and its chains. The Mid-Atlantic Ridge. The basement folding of the continent of Gondwana: marginal basement folds and internal virgation. Of the behavior of basement synclines   138

XXX. Glances (continuation). East Asia. Festoons of the continental margin. Lateral influence of the Indo-Angaran compression on East Asia; order of magnitude of the relative displacement of India and of Angara Land. Abyssal furrows, festoons, and marginal seas. Survival and deformation, in the system dismembered by stern stresses, of a part of the previous plan generated by bow stresses and foldings in general. Virgation of the Alaskides and virgation of the Philippines. Japan. Within sight of the Pacific   158

XXXI. Conclusion. The concept of generalized framed folding. Presentation of the concept. Of framed folding on the scale of cover folds, on the scale of new chains, on the scale of basement folding or continental scale. Hierarchical order of the scales and of the corresponding types of framed folding. This order is also, in a general way, that of the distributions of energy. The concept of generalized frame folding includes mobilism as a particular case. Of sima as frame, on the planetary scale. Value of mobilism. The common background of mobilism and of classical tectonics. Original mobilism and mobilism loaded with all the concrete tectonics. The notion of generalized framed folding transposes into concepts a vision of framed folding on all scales. This vision implies, in a state of extreme condensation, an approximate picture of the total deforming movement. It is independent of any theory. How it unfolds into innumerable fragmentary visions of the same character: images of movements or movements of images. How fragmentary visions are transposed into ideas. Flexibility of the behaviors and richness in operational procedures. No systematic induration. This general vision satisfies the requirement of synthesis without restricting the freedom of the mind with respect to theories. Advantages of this attitude. Theories and the superiority of concrete tectonics. Invention and criticism; visions and arrangement of ideas: the necessary equilibrium   160

Epilogue: Asia   165
Illustrations   166

QE
289
A7313
1977

APR 19 1978